Jens Fiedler

Radiative aspects in coupled nuclear fusion-fission processes

Jens Fiedler

Radiative aspects in coupled nuclear fusion-fission processes

The influence of heat and radiation heat conduction on the hydrodynamics of fission ignited fusion processes

Südwestdeutscher Verlag für Hochschulschriften

Impressum/Imprint (nur für Deutschland/ only for Germany)
Bibliografische Information der Deutschen Nationalbibliothek: Die Deutsche Nationalbibliothek verzeichnet diese Publikation in der Deutschen Nationalbibliografie; detaillierte bibliografische Daten sind im Internet über http://dnb.d-nb.de abrufbar.

Alle in diesem Buch genannten Marken und Produktnamen unterliegen warenzeichen-, markenoder patentrechtlichem Schutz bzw. sind Warenzeichen oder eingetragene Warenzeichen der jeweiligen Inhaber. Die Wiedergabe von Marken, Produktnamen, Gebrauchsnamen, Handelsnamen, Warenbezeichnungen u.s.w. in diesem Werk berechtigt auch ohne besondere Kennzeichnung nicht zu der Annahme, dass solche Namen im Sinne der Warenzeichen- und Markenschutzgesetzgebung als frei zu betrachten wären und daher von jedermann benutzt werden dürften.

Verlag: Südwestdeutscher Verlag für Hochschulschriften Aktiengesellschaft & Co. KG
Dudweiler Landstr. 99, 66123 Saarbrücken, Deutschland
Telefon +49 681 37 20 271-1, Telefax +49 681 37 20 271-0
Email: info@svh-verlag.de
Zugl.: Wuppertal, U, Diss., 2009

Herstellung in Deutschland:
Schaltungsdienst Lange o.H.G., Berlin
Books on Demand GmbH, Norderstedt
Reha GmbH, Saarbrücken
Amazon Distribution GmbH, Leipzig
ISBN: 978-3-8381-1746-1

Imprint (only for USA, GB)
Bibliographic information published by the Deutsche Nationalbibliothek: The Deutsche Nationalbibliothek lists this publication in the Deutsche Nationalbibliografie; detailed bibliographic data are available in the Internet at http://dnb.d-nb.de.

Any brand names and product names mentioned in this book are subject to trademark, brand or patent protection and are trademarks or registered trademarks of their respective holders. The use of brand names, product names, common names, trade names, product descriptions etc. even without a particular marking in this works is in no way to be construed to mean that such names may be regarded as unrestricted in respect of trademark and brand protection legislation and could thus be used by anyone.

Publisher: Südwestdeutscher Verlag für Hochschulschriften Aktiengesellschaft & Co. KG
Dudweiler Landstr. 99, 66123 Saarbrücken, Germany
Phone +49 681 37 20 271-1, Fax +49 681 37 20 271-0
Email: info@svh-verlag.de

Printed in the U.S.A.
Printed in the U.K. by (see last page)
ISBN: 978-3-8381-1746-1

Copyright © 2010 by the author and Südwestdeutscher Verlag für Hochschulschriften Aktiengesellschaft & Co. KG and licensors
All rights reserved. Saarbrücken 2010

Contents

Zusammenfassung	2
1 Motivation	**3**
2 Introduction	**5**
3 Physical model	**9**
3.1 Equations of radiation hydrodynamics	9
3.2 Radiative equilibrium and radiation energy	12
3.3 The one-dimensional spherical symmetric case	13
3.4 Equations of state	14
3.4.1 Equation of state for uranium/plutonium	14
3.4.2 Equation of state for deuterium/tritium	15
3.5 External energy contributions	22
3.5.1 Nuclear energy	22
3.5.2 Fusion model	23
4 Ionisation energies	**27**
4.1 Ionisation energies for hydrogen and helium	27
4.2 The model of Slater and its application on uranium and plutonium	27
4.3 Equilibrium of ionisation - the equation of Saha	32
4.4 Advanced methods	36
5 Radiation cross sections	**41**
5.1 Bremsstrahlung/Inverse bremsstrahlung	42
5.2 Total absorption coefficient	44
5.3 Effects of scattering of photons on electrons	45
5.4 Mean free path methods	46
5.4.1 Optical depth	46
5.4.2 Rosseland and Planck limit	47
6 Aspects of conduction	**53**
6.1 Heat conduction	53
6.2 Radiation diffusion approximation	57
6.2.1 Momentum equations	57
6.2.2 Radiation heat conduction - equilibrium diffusion approximation	59

CONTENTS

	6.2.3	Flux limiters	60
7	**Multigroup approach**		**63**
	7.1	Group integrated quantities	64
	7.2	Group integrated Planck function	65
	7.3	Absorption cross section	66
	7.4	Scattering cross section	67
	7.5	Scattering transfer cross section	69
	7.6	Moments of the group transfer scattering cross section	71
		7.6.1 Group transfer scattering cross section in the Thomson limit	71
		7.6.2 Frequency shift formula	72
		7.6.3 General case of the group transfer scattering cross section	73
		7.6.4 An asymptotic analytical solution	76
8	**Results**		**83**
	8.1	The coupled system without α-heating	85
	8.2	The coupled system including α-heating	89
	8.3	Summary	91
9	**Program System**		**101**
10	**Conclusion and Outlook**		**103**
A	**Appendix**		**105**
	A.1	Units	105
	A.2	Zeta-function	105
	A.3	Integration of Planck's function	105
	List of Figures		**110**
	List of Tables		**111**
	References		**119**

Zusammenfassung

Die Beurteilung von Sicherheitsrisiken, Schutz- und Dekontaminationsmaßnahmen im Hinblick auf nukleare Spaltanordnungen hängt maßgebend von der Fähigkeit ab, deren Wirkung beurteilen zu können. Fundierte Aussagen erfordern eine qualitative und quantitative Einschätzung des Zusammenwirkens der grundlegenden hydrodynamischen und nuklearen Vorgänge. Dafür sind keine detaillierten Kenntnisse über den Aufbau der nuklearen Explosionsanordnung notwendig. Hinreichende Modellannahmen liefern quantitative Abschätzung einer solchen Anordnung [Pritzker81].

Ein im Zusammenhang mit Spaltanordnungen stehendes Prinzip ist das der Fusionsverstärkung. Dabei werden durch den nuklearen Zündungsprozess zusätzliche Fusionsreaktionen angeregt. Die entstehenden Fusionsneutronen verstärken ihrerseits den Spaltprozess weiter. Dabei ist es das Ziel, den Abbrand des Spaltmaterials zu maximieren und die Menge des Spaltstoffes, bei gleichbleibender Energieausbeute, zu minimieren.

Der Prozess der fusionsverstärkten Kernspaltung wird durch eine Vielzahl von physikalischen Prozessen begleitet. Neben den Implosions- und Expansionsvorgängen unter Einwirkung von nuklearen Spalt- und Fusionsprozessen, beeinflusst die bei hohen Temperaturen entstehende Wärmestrahlung das System zusätzlich. Die Grundlage der theoretischen Behandlung eines solchen Systems bilden die hydrodynamischen Gleichungen, die die Bewegung von Flüssigkeiten und Gasen beschreiben. Die Wechselwirkung zwischen Materie, Spalt- und Strahlungsprozessen wird durch die Neutronen- und Strahlentransportgleichung modelliert. Die physikalische Nähe gestattet es, Modelle aus den Bereichen Plasmaphysik, Astrophysik, Trägheitsfusion und Reaktorphysik anzuwenden. Beispielsweise können die in der Astrophysik abgeleiteten Ausdrücke für die Opazitäten[1] entsprechend auf schwere Elemente angepasst und verwendet werden.

Mit Hilfe des Programmsystems **STEALTH-MCNP** wird die gegenseitige Beeinflussung zwischen veränderlichen Materialdichten und Teilchen- und Strahlungstransport numerisch untersucht. **STEALTH** [Chan78] simuliert den hydrodynamischen Verlauf der Modellanordnung und **MCNP** [X-5 Monte Carlo Team03] generiert die jeweiligen neutronenphysikalischen Größen.

---

[1]Materialabhängige Wirkungsquerschnitte, die für die Lösung der Strahlentransportproblematik benötigt werden.

CONTENTS

In der vorliegenden Dissertation wird erstmals der Einfluß der Wärmestrahlung in gekoppelten Fusion- Kernspaltungsanordnungen untersucht. Dabei gilt es, Abschätzungen für die Strahlenwirkungsquerschnitte zu finden, sowie den Strahlentransport hinreichend zu beschreiben. Sowohl für die Bestimmung der Wirkungsquerschnitte, die Behandlung des Strahlentransports als auch für die Beschreibung des Fusionsprozesse wurde die Annahme zugrunde gelegt, dass zu jedem Zeitpunkt und in jedem Raumpunkt ein lokales thermisches Gleichgewicht vorherrscht. Zu jedem Zeitpunkt und zu jedem Raumpunkt kann eine Temperatur zugeordnet werden. In diesem Fall vereinfachen sich die Verfahren zur Beschreibung der oben genannten physikalischen Prozesse maßgebend. Die Lösung der Strahlentransportgleichung wurde mit dem Ansatz der Strahlungswärmeleitung genähert.

Die neu betrachteten Aspekte sind die Beschreibung des Fusionsprozesses sowie die Einspeisung der Fusionsenergie in das gekoppelte System. Es wurde erstmals der elektronische Wärmetransports und die Strahlungswärmeleitung in einem kernspaltungsgetriebenen Fusionsplasma untersucht. Neben diesen numerischen Untersuchungen wurden für Streuterme, die in der Strahlentransportgleichung auftreten und die von Pritzker et al. [Pritzker76] numerisch untersucht wurden, approximative analytische Ausdrücke gefunden.

1 Motivation

The energy liberated by the sun results from fusion reactions. During those reactions light nuclei fuse to product nuclei. Since a long time it is known that the sum of masses of light nuclei exceeds the mass of the product nucleus. The difference in mass is related to energy by the formula of Einstein $E = \Delta mc^2$. The energy is released in form of kinetic energy to the reaction products. Fusion reactions occur when the species to be fused come very close to each other. They have to overcome Coulomb forces by their kinetic energy. Special conditions like high temperatures and densities are required to obtain a significant amount of energy by fusion reactions.

In astrophysical dimensions such conditions are fulfilled within the interiors of stars. The H-bomb first realized similar conditions on earth [Motz79]. Starting from the middle of the 20th century large scale experiments have been set up to realize fusion reactions of type

$$\text{deuterium} + \text{tritium} \rightarrow \text{helium-4} + \text{neutron} + 17.6 \text{ MeV}$$

in laboratory dimensions. At very short distances and very short time scales the inertial confinement fusion (ICF) is a powerful tool to successfully initialise a nuclear fusion. Here, the surface of a deuterium-tritium (D-T) capsule (diameter $\approx$ 1 mm) is ignited by high power short pulsed lasers. The absorbed energy gets transported to the fuel by shockwaves and by the mechanism of electron thermal and radiation heat conduction depending on temperature.

A concept very similar to that of ICF is the configuration of fusion boosted fission systems. See figure (2). Hereby fission processes are enhanced by additional neutrons which are produced by fusion reactions. The D-T core (diameter $\approx$ 70 mm) is surrounded by a shell of fissile material which is surrounded by an amount of explosive. The nuclear energy release by the fissile material depends on its compression. Fusion conditions occur when a significant amount of fission energy is liberated by the fissile material and D-T plasma is highly compressed and heated. Fusion neutrons produced by those reactions stimulate likewise the fission processes.

The aim of this thesis is to study fusion plasma in dimension of around several centimetres surrounded by a fissile material. At high temperatures a significant amount of energy is governed by radiation. Radiation and the encapsulated material generally affect each other.

1 Motivation

The main aspect is the investigation of the electron thermal and radiation heat conduction processes coupled to the hydrodynamical behaviour of hot and dense D-T and fissile plasma under condition described above. It is of interest to discover how far conduction plays a role within the burn-up of a D-T plasma ignited by fission processes.

2 Introduction

Starting from the determination of stellar atmospheres in the early decades of the 20th century the physics of radiation transport has been starting to become a powerful tool for understanding of the physical processes in stellar envelopes and interiors of stars. Radiation transport means the energy redistribution within a medium by emission and absorption of photons [Apruzese02]. The measurement of line absorption spectra emitted by stellar systems results in the knowledge of underlying elements existent in that system.

Nowadays, the absorption and emission contribution of different materials are essential to setup efficient inertial confinement fusion facilities. In contrast to the direct driven setup the principle of indirect concept of inertial driven confinement fusion [Atzeni87, Vehn97, McCrory08] is to convert high power laser or heavy ion beams to Hohlraum generated X-rays. See figures (2) and (1). Such a generated radiation field is symmetric and offers a highly symmetric implosion of the fusion capsule [Tahir97]. The main idea is to reduce the influence of oscillations in laser beams, which results in instabilities while compressing the capsule. It is therefore of practical interest to find materials or mixtures of materials with a high reemission and X-ray conversion efficiency [Yan02]. However, the disadvantage of the indirect driven technology is the energy loss by converting laser or heavy ion energy to X-rays.

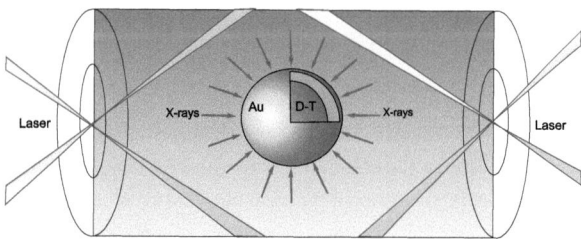

Figure 1: Principle of an indirect driven inertial confinement facility. The laser or particle beams are converted to X-rays by the Hohlraum walls.

A lot of physical models have been developed to investigate the absorption and emission behaviour of elements. Of widely use are models assuming a local thermal equilibrium (LTE) [Armstrong72, Zel'dovich66, Pritzker75, Pritzker76]. In that case at every

2 INTRODUCTION

time in every spatial coordinate of the material a temperature is defined. For plasma in LTE conditions the different ionisation stages of matter are determined by solving the Saha-Boltzmann equation, please refer to section (4) on page 27 additionally. In local thermal equilibrium the absorption and emission coefficient are connected by the rule of [Armstrong72, Kourganoff63]. The increase of computational power gives contribution to activities in opacity calculations in non-local thermal equilibrium conditions by solving the rate equations [Yan02, Zeqing06]. The influence of degeneration of an electron plasma in opacity calculations has been discussed in [Khalfaoui97].

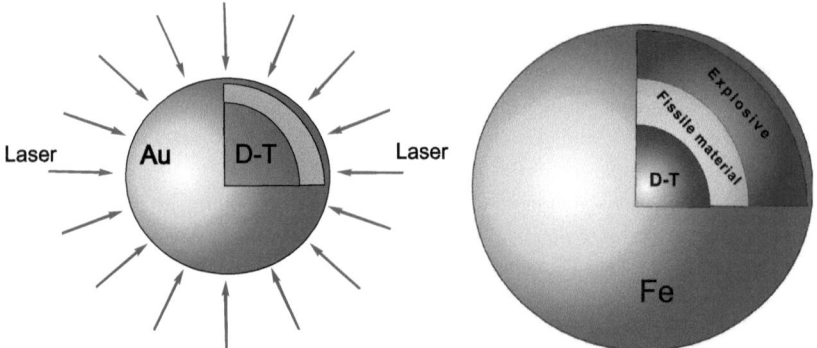

Figure 2: Principle of a direct driven inertial confinement (ICF) facility (left picture) and of a fusion boosted nuclear system. The diameter of a capsule of an ICF configuration is around 1 mm and of a fusion boosted system around 70 mm.

The radiation field and the encapsulated material generally affect each other. The description of such a model requires the solution of the radiation transport equation, which is a complex task. Analytically, the coupled problem of radiation transport and hydrodynamics in a shock wave regime was solved approximately by [Marshak58]. Radiation transport is studied widely in ICF research as well as in stellar atmospheres [Chandrasekhar60]. Neutron transport phenomena are well discussed in reactor physics [Bell70, Kourganoff63], but only a handful results are published which deal with the combined problem of neutron- and radiation transport [Pritzker81].

In the late 70ties of the last century Pritzker et al. [Pritzker81] have developed a program system called **SHYRAN** which allows studying the physics of interactions between hydrodynamics, fission and radiation transport. This program solves a set of finite difference equations in one-dimensional spherical Lagrangian coordinates, coupled with multigroup S-4 neutronics[2], multigroup gamma diffusion and one-group thermal radiation diffusion in a discrete time mesh. Their published results are based on investigating the behaviour of an idealised imploding nuclear fission device presented by an one-dimensional spherical geometry.

For the first time the concept of a fusion boosted system at which fusion material is concentrated in a sphere surrounded by fissionable material is investigated in this thesis. For this purpose a coupling between the hydrodynamic program STEALTH [Chan78] and the semi-timedependent neutron transport program system MCNP [X-5 Monte Carlo Team03] has been developed. Moreover STEALTH was significantly extended for depositing nuclear and fusion energy to the physical system, calculation of the thermonuclear burn-up and for computing conduction problems. To make oneself familiar with the hydrodynamical equations section (3) acts as a theoretical introduction to this matter. Therein models for calculating the fusion energy, thermonuclear burn-up and fission energy contributions are presented. The microscopic and macroscopic radiation transport cross sections at very high temperatures (several keV) and high densities (several orders of solid state density) for hydrogen and for heavy elements like uranium and plutonium are investigated. Those radiation coefficients depend on temperature, electron density as well as on the level of ionisation. In section (4) the model of Slater [Slater30] for a rough estimation of the ionisation potentials is presented. The average ionisation stage and the electron density are obtained by solving the equation of Saha [Landau87] for uranium and plutonium. In section (5) the radiation cross sections are introduced. Therein expressions for freefree, total absorption and photon scattering cross sections are given. The theoretical results have been applied to hydrogen and uranium. The following section (6) deals with the electronic and radiation heat conduction approximations. Those approximations give the possibility to study radiation transport effects in lowest order of anisotropy. Based on the radiation cross sections expressions for the radiation heat conduction coefficients are derived. Beyond the concept of conduction an analytical method for calculating the multigroup radiation cross section required for solving the multigroup equation of radiative transfer will be explained and presented in section (7). Finally, results from coupled fissionfusion calculations taking zeroth order radiation transport equation into account are given in section (8).

---

[2]S-N approximation: Discrete ordinate approximation to N-th order.

3 Physical model

Theoretically, the present model is described by a set of coupled partial (integro-) differential equations. The flow of particles is given by the solution of the conservation equations for mass, momentum and energy. Nuclear and radiation processes are determined by the neutron and radiation transport equation. Moreover, the equation of state for a deuterium-tritium and fissile material plasma as well as a description of the fusion process are required. In this section the basic formulas describing the underlying physical processes and approximations to reduce the numerical and technical effort are introduced.

3.1 Equations of radiation hydrodynamics

The basic equations of hydrodynamics consist of the equation of continuity, the momentum equation and the energy equation [Zel'dovich66]. The equation of continuity reads

$$\frac{\partial \rho}{\partial t} + \nabla \cdot (\rho \mathbf{u}) = 0 \tag{3.1.1}$$

where $\mathbf{u}$ is the collective velocity of the fluid, ρ the density and t the time. The momentum equation is given by

$$\frac{\partial}{\partial t}(\rho \mathbf{u} + \mathbf{G}) + \nabla \cdot \left(\hat{\mathbf{\Pi}} + \hat{\mathbf{p}} \right) = 0 \tag{3.1.2}$$

where the terms $\mathbf{G}$, $\hat{\mathbf{\Pi}}$ and $\hat{\mathbf{p}}$ are the expressions for the momentum density, momentum tensor and the momentum tensor of the radiation flux. Definitions will be given below. The equation of energy including contributions by radiation reads

$$\frac{\partial}{\partial t}\left(\rho e + \frac{\rho \mathbf{u}^2}{2} + U_{rad}\right) + \nabla \cdot \left(\mathbf{u}\left(\rho e + p + \frac{\rho \mathbf{u}^2}{2}\right) + \mathbf{S}\right) = \underbrace{\epsilon_1 Q_f + \epsilon_2 R}_{\text{external energy source}} \tag{3.1.3}$$

U_{rad} is the radiation energy density, p the pressure, $\mathbf{S}$ the radiation energy flux and e the energy per mass contribution of the fluid. The sum $\epsilon_1 Q_f + \epsilon_2 R$ on the right hand side denotes the external energy source and consists of the contributions from nuclear fission and fusion processes. Q_f is the fission rate density and R is the fusion rate density. These quantities are defined in section (3.5). $\epsilon_1 \approx 180$ MeV is the energy which is delivered by one fission process. $\epsilon_2 \approx 3.5$ MeV is the energy which is delivered by one fusion process. We restrict ourselves to α-heating. That means, only the energy from α-particles is carried

3 Physical Model

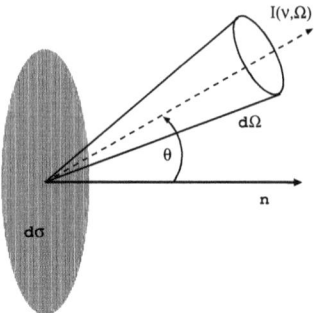

Figure 3: Definition of the specific intensity [Pomraning73].

to the system. The neutron energy contributions will be neglected.

Suppose that the photons in a radiation field are distributed by a function $f_\nu(\mathbf{r}, \mathbf{\Omega}, t)$. ν means the photon frequency, $\mathbf{\Omega}$ the direction of travel of photons, $\mathbf{r}$ the spatial coordinate and t the time. Each photon has an energy $E = h\nu$. The amount of energy dE transported by photons within a frequency interval $d\nu$, travelling in direction $\mathbf{\Omega}$ in time interval dt and solid angle element $d\mathbf{\Omega}$ through an area $d\sigma$ is given by [Pomraning73, Mihalas78]. Refer to figure (3).

$$dE = ch\nu f_\nu(\mathbf{r}, \mathbf{\Omega}, t) \cos\theta \, d\nu \, d\mathbf{\Omega} \, d\sigma \, dt. \quad (3.1.4)$$

$\cos\theta$ is the angle of the normal vector of area $d\sigma$ with $\mathbf{\Omega}$. Considering (3.1.4) the specific intensity $I_\nu(\mathbf{r}, \mathbf{\Omega}, t)$ is defined by

$$I_\nu(\mathbf{r}, \mathbf{\Omega}, t) = ch\nu f_\nu(\mathbf{r}, \mathbf{\Omega}, t). \quad (3.1.5)$$

The specific intensity is a solution of the equation of radiative transfer [Pomraning73]

$$\frac{1}{c}\frac{\partial}{\partial t}I_\nu + \mathbf{\Omega} \cdot \nabla I_\nu = \mathcal{J}_\nu(\mathbf{r}, \mathbf{\Omega}, t)\left(1 + \frac{c^2 I_\nu}{2h\nu^3}\right) - \Sigma_a(\nu)I_\nu + \mathcal{C}\left[I_\nu\right] \quad (3.1.6)$$

where Σ_a is the macroscopic absorption cross section, $\mathcal{J}_\nu$ the emission source term due to spontaneous emission and $\mathcal{C}$ the collision operator

3.1 Equations of Radiation Hydrodynamics

$$\mathcal{C}[I_\nu] = \int_0^\infty d\nu' \int_{4\pi} d\Omega' \frac{\nu}{\nu'} \Sigma_s \left(\nu' \to \nu, \Omega' \cdot \Omega\right) I_{\nu'} \left(1 + \frac{c^2 I_\nu}{2h\nu^3}\right) \\ - \int_0^\infty d\nu' \int_{4\pi} d\Omega' \Sigma_s \left(\nu \to \nu', \Omega \cdot \Omega'\right) I_\nu \left(1 + \frac{c^2 I_{\nu'}}{2h\nu'^3}\right). \quad (3.1.7)$$

$\Sigma_s \left(\nu' \to \nu, \Omega' \cdot \Omega\right)$ is the differential scattering cross section, which is defined in such a way, that $\Sigma_s \left(\nu' \to \nu, \Omega' \cdot \Omega\right) d\nu \, d\Omega \, ds$ means the probability of a photon being scattered from frequency ν' to frequency ν contained in the frequency interval $d\nu$, and from Ω' to Ω contained in $d\Omega$ in travelling a distance ds [Pomraning73]. Let be ν and Ω the frequency and direction of interest. In upper integral in (3.1.7) sums all inscattering contributions of photons from all frequencies ν' and directions Ω' to ν and Ω. The lower integral of (3.1.7) counts all outscattering contributions of photons from ν and Ω to all frequencies ν' and directions Ω'. The increase of the probabilities of emission and scattering by the factors $\left(1 + \frac{c^2 I_\nu}{2h\nu^3}\right)$ and $\left(1 + \frac{c^2 I_{\nu'}}{2h\nu'^3}\right)$ in (3.1.6) and (3.1.7) results from the bosonic property of photons. That means a number of photons belongs already to the final state following the interaction. Generally this is called *induced emission* [Pomraning73].

The coefficients appearing in (3.1.6) and (3.1.7) describe the interaction of photons with the encapsule material.

Pressure, energy density and flux of the thermal radiation in a high temperature plasma are determined by the specific intensity. The radiation energy density is defined by

$$U_{rad} = \int_0^\infty d\nu \, U_\nu = \frac{1}{c} \int_0^\infty d\nu \int_{4\pi} d\Omega \, I_\nu \left(\mathbf{r}, \Omega, t\right) \quad (3.1.8)$$

the radiation energy flux reads

$$\mathbf{S} = \int_0^\infty d\nu \, \mathbf{S}_\nu = \int_0^\infty d\nu \int_{4\pi} d\Omega \, \Omega I_\nu \left(\mathbf{r}, \Omega, t\right) \quad (3.1.9)$$

and the momentum tensor of the radiation flux is introduced by

$$\hat{\mathbf{p}} = \int_0^\infty d\nu \, \hat{\mathbf{p}}_\nu = \frac{1}{c} \int_0^\infty d\nu \int_{4\pi} d\Omega \, \Omega \Omega I_\nu \left(\mathbf{r}, \Omega, t\right). \quad (3.1.10)$$

3 Physical model

3.2 Radiative equilibrium and radiation energy

It is cost intensive to solve the equation of radiation transfer accurately due to its complexity. Most often the emission term $\mathcal{J}_\nu$ is expressed by the absorption coefficient and taking the source function to be Planckian. In detail that means, one introduces the source function by [Kourganoff63]

$$S_\nu\left(\mathbf{r},\mathbf{\Omega},t\right) = \frac{\mathcal{J}_\nu}{\Sigma_a(\nu)}\left(1 + \frac{c^2 I_\nu}{2h\nu^3}\right), \quad (3.2.1)$$

and sets $S_\nu\left(\mathbf{r},\mathbf{\Omega},t\right) = B_\nu(\theta(\mathbf{r},t))$, $\theta = k_B T$. That is to say, that at every spatial coordinate and at every time step a local thermal equilibrium is assumed. In local thermal equilibrium the source function is given by Planck's function

$$B_\nu(\theta) = \frac{2h\nu^3}{c^2}\left(\exp\left(h\nu/\theta\right) - 1\right)^{-1}. \quad (3.2.2)$$

In that way one introduces

$$B_\nu(\theta)\Sigma_a(\nu) = \mathcal{J}_\nu\left(1 + \frac{c^2 B_\nu}{2h\nu^3}\right). \quad (3.2.3)$$

Inserting Planck's function (3.2.2) results in

$$B_\nu(\theta)\Sigma_a(\nu)\left(1 - \exp(-h\nu/\theta)\right) = \mathcal{J}_\nu, \quad (3.2.4)$$

which is the rule of Kirchhoff [Kourganoff63]. The term $\Sigma'_a(\nu) = \Sigma_a(\nu)\left(1 - \exp(-h\nu/\theta)\right)$ is the absorption coefficient corrected by induced emission. Using the results leads to a modified equation of radiation transfer

$$\frac{1}{c}\frac{\partial}{\partial t}I_\nu + \mathbf{\Omega}\cdot\nabla I_\nu = \Sigma'_a(\nu)\left(B_\nu(\theta) - I_\nu\right) + \mathcal{C}\left[I_\nu\right]. \quad (3.2.5)$$

That equation is used within this thesis. Solving of equation (3.2.5) is still numerically and technically cost intensive. Some approximate solutions will be discussed in section (6).

The radiation energy density used within this thesis is assumed to be in radiative equilibrium. In that case U_ν is given by the Stefan Boltzmann law

$$U_\nu = \frac{4\pi}{c}B_\nu(\theta) \quad (3.2.6)$$

which leads by frequency integration over the complete spectrum to its usual form

$$U_{rad} = a_{rad}T^4, \tag{3.2.7}$$

where a_{rad} is the radiation constant. A detailed calculation is given in the appendix. The radiation pressure is often approximated by

$$p_{rad} = \frac{U_{rad}}{3} = \frac{a_{rad}}{3}T^4, \tag{3.2.8}$$

which implies an isotropic behaviour.

3.3 The one-dimensional spherical symmetric case

The underlying physical problem is of spherical symmetry. That problem is solved in one dimension. In that case the equation of continuity (3.1.1) is simplified to

$$\frac{\partial \rho}{\partial t} + \frac{1}{r^2}\frac{\partial}{\partial r^2}\left(\rho r^2 u_r\right) = 0. \tag{3.3.1}$$

The momentum equation (3.1.2) becomes

$$\frac{\partial(\rho u_r)}{\partial t} + \frac{1}{r^2}\frac{\partial(\rho r^2 u_r^2)}{\partial r} = -\frac{1}{\rho}\frac{\partial}{\partial r}(p + p_{rad}) \tag{3.3.2}$$

G in (3.1.2) is assumed to be of small effect and will be neglected [Zel'dovich66] for the present case. The energy equation (3.1.3) reads

$$\frac{\partial}{\partial t}\left(\rho e + \frac{\rho u^2}{2} + U_{rad}\right) + \frac{1}{r^2}\frac{\partial}{\partial r}\left(ur^2\left(\rho e + p + \frac{\rho u^2}{2}\right) + r^2 S_r\right) = \epsilon_1 Q_f + \epsilon_2 R, \tag{3.3.3}$$

where S_r is the radial component of the radiation flux. The radiation energy density U_{rad} is given by (3.2.7). The radiation pressure p_{rad} is assumed to be isotropic and given by (3.2.8). The hydrodynamical equations are transformed further to Lagrangian coordinates and subsequently solved numerically by the program **STEALTH**. The external energy contributions as well as the radiation energy contributions will be discussed below. As one can see by equations (3.3.1), (3.3.2) and (3.3.3) and setting Q_f, R, U_{rad}, S_r and p_{rad} equal to zero for a moment, there is one additionally relation missing to have a closed system of equations. This required relation comes from thermodynamics and connects pressure, density and internal energy and is called equation of state. For the present problem one has to find adequate approximations for fission material and for the deuterium-tritium plasma.

3 Physical model

3.4 Equations of state

3.4.1 Equation of state for uranium/plutonium

For the current investigations a three-term equation of state for uranium and plutonium is proposed [Zel'dovich66]

$$p(\mu, T) = p_c(\mu) + G(\mu)c_V\rho_0(\mu+1)T + \frac{1}{2}G_e(\mu,T)\beta_0\rho_0(\mu+1)^{1-G_e(\mu,T)}T^2 \quad (3.4.1)$$

$$e(\mu, T) = e_c(\mu) + c_V T + \frac{1}{2}\beta_0(\mu+1)^{-G_e(\mu,T)}T^2 \quad (3.4.2)$$

where p_c is the cold pressure, e_c the cold energy, G is called the Grüneisen coefficient, G_e the electronic Grüneisen coefficient, T the temperature, ρ the material density, ρ_0 the material density at some standard conditions, c_V the specific heat capacity at constant volume, β_0 the specific electron heat coefficient and the compression $\mu = (\rho/\rho_0) - 1$. The constants in (3.4.1) and (3.4.2) have been studied widely by [Hafner91]. Therein the coefficients for uranium are given by

$$\rho_0 = 18.9 \, \frac{\text{g}}{\text{cm}^3}, \quad c_V \approx 1.05 \times 10^{-6} \, \frac{10^{12}\text{erg}}{\text{g K}}, \quad \beta_0 \approx 1.75 \times 10^{-10} \, \frac{10^{12}\text{erg}}{\text{g K}^2}.$$

The electronic Grüneisen coefficient is a slowly varying function in a wide range of μ and T and therefore set to be constant, $G_e \approx 0.5$. The expressions for the cold pressure, cold energy and the Grüneisen coefficient read [Hafner91, Zel'dovich66]

$$p_c = \mu(\mu+1)^\phi \quad (3.4.3)$$

$$e_c = \frac{1}{\rho_0}\int_0^\mu \frac{p_c(\mu')}{(\mu'+1)^2}d\mu' \quad (3.4.4)$$

$$G = \frac{2}{3} + \frac{\mu+1}{2}\frac{d}{d\mu}\ln\frac{d}{d\mu}\frac{p_c(\mu)}{(\mu+1)^{2/3}} \quad (3.4.5)$$

where

$$\phi = 2/3 + 1/\psi \quad (3.4.6)$$

$$\psi \approx 0.561 + 0.0538\ln(\mu^2 + \sqrt{\mu^4+1}). \quad (3.4.7)$$

3.4 Equations of state

The three-term equation of state for uranium is obtained by inserting (3.4.2) in (3.4.1). In this way one finds

$$p(\rho,T) = p_c(\mu) + \frac{\rho_0}{2}(\mu+1)\left[e - e_c(\mu) + (2G(\mu)-1)c_V T\right] \qquad (3.4.8)$$

An expression for the temperature is found by solving (3.4.2) for T

$$T = \frac{c_V \sqrt{\mu+1}}{\beta_0}\left(\sqrt{1+x} - 1\right) \qquad (3.4.9)$$

where

$$x = \frac{2\beta_0\left(e_c(\rho) - e\right)}{c_V^2 \sqrt{\mu+1}}. \qquad (3.4.10)$$

Applying the first law of thermodynamics [Landau87]

$$dQ = de + pd\left(\frac{1}{\rho}\right) = \left(\frac{\partial e}{\partial T}\right)dT + \left(\frac{\partial e}{\partial \rho}\right)d\rho + pd\left(\frac{1}{\rho}\right), \qquad (3.4.11)$$

the specific heat depending on compression and temperature reads

$$c_V(T,\mu) = \left(\frac{dQ}{dT}\right)_\rho = c_V + \beta_0 \sqrt{\frac{1}{1+\mu}} T. \qquad (3.4.12)$$

For numerical purposes the squared sound speed is calculated. Taking the definition $v_s^2 = \left(\frac{\partial p}{\partial \rho}\right)_s$ the calculation by using Gibbs relation gives [Hafner91]

$$v_s^2 = \frac{1}{\rho_0}\left[\frac{\partial p}{\partial \mu} + \frac{p}{2}\frac{1}{\mu+1}\left(1 + \frac{2G(\mu)-1}{\sqrt{1+x}}\right)\right]. \qquad (3.4.13)$$

The partial derivative $\partial p/\partial \mu$ has been evaluated numerically. Other contributions like the electrostatic interaction, the ionisation- and excitation energy and the radiation energy will be part of future developing processes. Such contributions have been considered by Pritzker et al. [Pritzker71].

3.4.2 Equation of state for deuterium/tritium

The equation of state for the fusion material in the present model is realized by an equation of state for an ideal gas. Comparing the notes of [Pai66], the radiation energy contribution is added to the internal energy of the fusion material. In the present case the internal energy density U is represented by

3 Physical model

$$U = c_V \rho T + a_{rad} T^4, \tag{3.4.14}$$

where c_V is the specific heat capacity at constant volume, ρ is the density, T the temperature and a_{rad} the radiation constant. The second contribution in (3.4.14) is the radiation energy density in radiative equilibrium. By noting $U = e\rho$ equation (3.4.14) is written as

$$e = c_V T + \frac{a_{rad} T^4}{\rho}, \tag{3.4.15}$$

where e is the energy per mass. In the present model the value of c_V has been determined by

$$c_V = \frac{1}{\gamma - 1} \frac{k_B}{m_N}. \tag{3.4.16}$$

On assumption that the D-T plasma is immediately fully ionised m_N means the averaged fusion particle mass $m_N = (m_{DT} + 2m_e)/4 = 2.088 \times 10^{-24}$ g, where m_{DT} is the sum of the mass of a deuterium ion and of a tritium ion. The adiabatic coefficient γ depends on the degrees of freedom of gas particles by

$$\gamma = \frac{c_V}{c_p} = \frac{f+2}{f}. \tag{3.4.17}$$

Particles in an ideal gas have three degrees of freedom f in translation, hence $\gamma = 5/3$. Because of the radiation contribution at high temperatures one has to include three additional degrees of freedom for radiation, thus $\gamma = 4/3$. The approximation $\gamma = 1.4$ has been proposed by Hafner [Hafner09] taking radiation contributions for an ideal gas equation of state for the D-T plasma at high temperatures into account. The specific heats for the D-T plasma used within the program system are given in table (1).

γ	c_V [10^{12} erg/g/K]
1.4	1.653×10^{-4}
1.667	9.918×10^{-5}

Table 1: Specific heats at constant volume for a D-T plasma.

In case of radiation in equilibrium and isotropic radiation pressure the pressure p of the D-T plasma reads

$$p = (\gamma - 1) c_V \rho T + \frac{a_{rad}}{3} T^4 = p_g + p_{rad}. \tag{3.4.18}$$

3.4 EQUATIONS OF STATE

In the current model the D-T plasma is divided into several cells. Each cell gets its specific temperature in case of a local thermal equilibrium by a relation between the hydrodynamical quantities. Depending from that temperature the radiation pressure and energy are calculated. Density, temperature and pressure of an ideal gas are connected by the following relations [Landau87]

$$\frac{dp}{p} - \gamma_1 \frac{d\rho}{\rho} = 0 \qquad (3.4.19)$$

$$\frac{dp}{p} + \frac{\gamma_2}{1-\gamma_2}\frac{dT}{T} = 0 \qquad (3.4.20)$$

$$\frac{dT}{T} - (\gamma_3 - 1)\frac{d\rho}{\rho} = 0. \qquad (3.4.21)$$

These relations are known as law of Boyle-Mariotte (3.4.19), law of Amontons (3.4.20) and law of Gay-Lussac (3.4.21). Using the above relations one finds the connection

$$\frac{\gamma_1}{1+\gamma_3} = \frac{\gamma_2}{1-\gamma_2} \qquad (3.4.22)$$

between the coefficients γ_1, γ_2 and γ_3. The differential of pressure is

$$dp = \left(\frac{4}{3}a_{rad}T^4 + (\gamma-1)c_V\rho T\right)\frac{dT}{T} + ((\gamma-1)c_V\rho T)\frac{d\rho}{\rho}. \qquad (3.4.23)$$

Inserting (3.4.23) in (3.4.19) and comparing to (3.4.11), where in adiabatic case $dQ = 0$ is valid, one finds γ_1.

$$\gamma_1 = \frac{(4-3b)^2(\gamma-1)}{12(1-b)(\gamma-1)+b}, \qquad (3.4.24)$$

where $b = p_g/p$. Similar one finds γ_2 by using (3.4.19) in (3.4.21).

$$\gamma_2 = \frac{\gamma_1(4-3b)}{3(1-b)\gamma_1 + b} \qquad (3.4.25)$$

γ_3 is obtained by help of (3.4.22).

$$\gamma_3 = 1 + \frac{\gamma_1 - b}{4 - 3b} \qquad (3.4.26)$$

When the radiation effects become dominant the above relations behave like an ideal gas with $\gamma = 4/3$. Figure (4) shows the different energy contribution to the internal energy depending from temperature. The equilibrium radiation energy term appearing in

3 Physical model

(3.4.14) becomes the main contribution to the internal energy density at temperatures above 10 keV. Figure (5) presents the pressure depending from internal energy density for different particle densities. With increasing internal energy density the pressure converges to radiation pressure given by (3.2.8). The behaviour of the adiabatic coefficients γ_1, γ_2 and γ_3 is shown in figure (6). At low temperatures ($k_B T \leq 3$ keV) the behaviour is that of an ideal gas taking $\gamma_1 = \gamma_2 = \gamma_3 = \gamma = 5/3$ whereas at very high temperatures ($k_B T \geq 70$ keV) the behaviour is that of an ideal gas taking $\gamma_1 = \gamma_2 = \gamma_3 = \gamma = 4/3$.

The specific heats are transformed to

$$c_{Vr} = \left(\frac{dQ}{dT}\right)_\rho = \frac{c_V}{b}\left(12(1-b)(\gamma-1)+b\right) \tag{3.4.27}$$

$$c_{pr} = \left(\frac{dQ}{dT}\right)_p = c_{Vr} - T\frac{\left(\frac{\partial p}{\partial T}\right)_V^2}{\left(\frac{\partial p}{\partial V}\right)_T} = \frac{c_V}{b^2}\gamma_1\left(12(1-b)(\gamma-1)+b\right). \tag{3.4.28}$$

The squared sound speed reads

$$v_s^2 = (\gamma-1)c_V T + \frac{(\gamma-1)c_V T + 4/3 a_{rad} T^3}{4 a_{rad} T^3 + \rho c_V}\left(\frac{a_{rad} T^4 + p}{\rho}\right) = \gamma_1 \frac{p}{\rho}. \tag{3.4.29}$$

3.4 EQUATIONS OF STATE

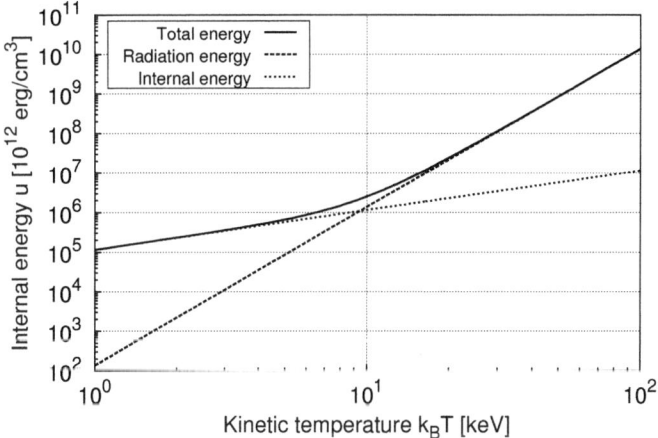

Figure 4: The dependency of the internal energy density of a D-T plasma on the temperature at a density $\rho = 100$ g/cm^3. At high temperatures ($k_B T \geq 10$ keV) the radiation energy becomes the dominant part of the total energy.

3 Physical model

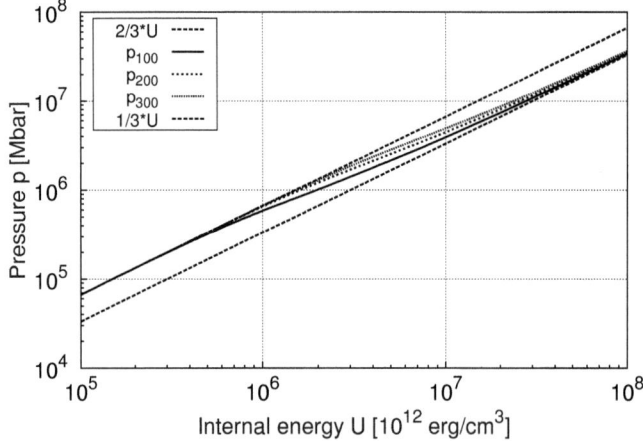

Figure 5: Internal energy vs. pressure in a deuterium-tritium plasma at different densities. p_{100} means pressure measured with density $\rho = 100$ g/cm^3, p_{200} with density $\rho = 200$ g/cm^3 and p_{300} with density $\rho = 300$ g/cm^3, respectively. U means the internal energy. The limit $U/3$ is the radiation pressure in radiative equilibrium (Stefan Boltzmann law). Depending on density the pressure converges to radiation pressure as valid in radiative equilibrium with increasing internal energy. On the other hand, the pressure behaves like the pressure from the ideal gas law in the limit of low internal energies, that means at low temperatures. In that case the radiation contribution does not play any role.

3.4 EQUATIONS OF STATE

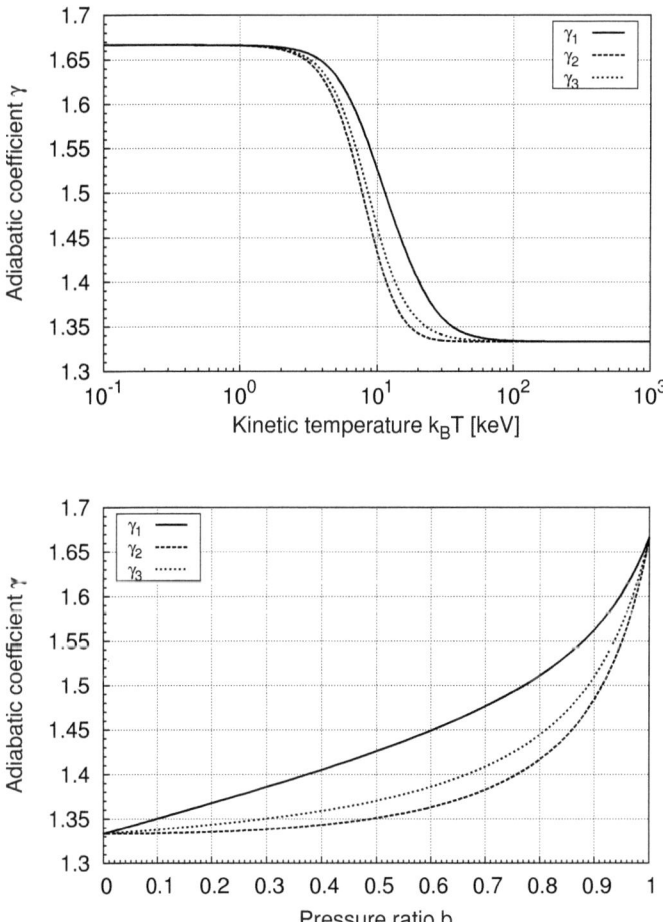

Figure 6: Behaviour of the adiabatic coefficients depending on temperature and on the pressure ratio $b = p_g/(p_g + p_{rad})$. In the limit of $b = 0$ and high temperatures the ideal gas behaves like a gas with $\gamma = 4/3$.

3 Physical model

3.5 External energy contributions

The energy equation (3.1.3) has been extended for external terms of fission and fusion energy. Formally, those supplements are calculated by solving the neutron transport equation and fusion reaction rates.

3.5.1 Nuclear energy

The release of nuclear energy is determined by the number of fissions in a fissile material. The number of fissions is affected by the interactions of neutrons with fissile material. Those mechanisms are well described by the neutron transport equation. Let $\mathbf{r}$ be the spatial component, t the time, $\mathbf{\Omega}$ the neutron flow direction, E the neutron energy, $v = v(E)$ the neutron velocity and $\phi = \phi(\mathbf{r}, \mathbf{\Omega}, E, t)$ or $\phi' = \phi(\mathbf{r}, \mathbf{\Omega}', E', t)$ the directional neutron flux. Then, the formal neutron transport equation is given by [Bell70]

$$\frac{1}{v}\frac{\partial}{\partial t}\phi + \mathbf{\Omega} \cdot \nabla\phi = -\sum_x \Sigma_x(\mathbf{r}, E, t)\phi \\ + \int_0^\infty dE' \int_{4\pi} d\mathbf{\Omega}' \sum_x \Sigma_x(\mathbf{r}, E', t) P_x(\mathbf{r}, \mathbf{\Omega}' \to \mathbf{\Omega}, E' \to E)\phi'. \quad (3.5.1)$$

$\Sigma_x(\mathbf{r}, E, t)$ or $\Sigma_x(\mathbf{r}, E', t)$ is the macroscopic cross section for a process x, where x involves s-cattering, f-ission, a-bsorption, ... P_x determines the transition probability for process x. The number of fissions is given by

$$Q_f(\mathbf{r}, \mathbf{\Omega}, E, t) = \Sigma_f(\mathbf{r}, E, t)\phi(\mathbf{r}, \mathbf{\Omega}, E, t). \quad (3.5.2)$$

Σ_f represents the macroscopic fission cross section. By an angle integration of (3.5.2) and defining the angle integrated neutron flux

$$\Phi(\mathbf{r}, E, t) = \int_{4\pi} d\mathbf{\Omega}\ \phi(\mathbf{r}, \mathbf{\Omega}, E, t)$$

followed by an integration over the energy spectrum one obtains the fission rate density

$$Q_f(\mathbf{r}, t) = \int_0^\infty dE\ \Sigma_f(\mathbf{r}, E, t)\Phi(\mathbf{r}, E, t). \quad (3.5.3)$$

Solving (3.5.1) for ϕ leads to the knowledge of the fission rate density Q_f. Due to the complicated mathematical nature of (3.5.1) ϕ and Q_f are obtained numerically by the program MCNP [X-5 Monte Carlo Team03].

3.5 External energy contributions

The nuclear burn-up is estimated by determination of the ratio

$$\frac{\text{Number of fissions}}{\text{Number of initial heavy metal atoms}}. \tag{3.5.4}$$

The nuclear burn-up depends on fission processes and on the hydrodynamic flow. Transforming to a Lagrange coordinate system the ratio (3.5.4) is determined by fission processes only. The burn-up isotopes are not calculated at the moment, but this is an aim for future investigations.

3.5.2 Fusion model

Given ion densities n_A and n_B of species A and B, the fusion rate per volume of the participating species a and b is [Duderstadt82]

$$R(r,t) = n_A(r,t) n_B(r,t) \langle v\sigma_{AB}\rangle (r,t), \tag{3.5.5}$$

where $\langle v\sigma_{AB}\rangle$ is the velocity averaged fusion rate. By assumption of a local thermal equilibrium the cross section depends on temperature only. Hence, $\langle v\sigma_{AB}\rangle (r,t) = \langle v\sigma_{AB}\rangle (T(r,t))$. In that case the velocity distribution is given by a Maxwell-Boltzmann distribution. The D-T fusion reaction

$$^2_1\text{D} + ^3_1\text{T} \rightarrow ^4_2\text{He } (3.5 \text{ MeV}) + ^1_0\text{n } (14.1 \text{ MeV})$$

received the most attention in fusion research because the reaction proceeds at a rate almost two orders of magnitude larger than the following D-D fusion for low temperatures. Therefore the case `A = D-euterium` and `B = T-ritium` is considered only. The fusion cross sections depending on temperature for D-T reactions has been provided in [Duderstadt82]. The data therein have been prepared for numerical fittings. In this way the D-T cross section is approximated by

$$\begin{aligned}\langle v\sigma_{DT}\rangle (k_B T) = \langle v\sigma_{DT}\rangle_0 \exp\Big[&\zeta \ln^2(k_B T/k_B T_0) \\ &+ \iota \ln^3(k_B T/k_B T_0) + \tau \ln^4(k_B T/k_B T_0)\Big]\end{aligned} \tag{3.5.6}$$

where the constants ζ, ι and τ are given in table (2).

3 Physical model

$\langle v\sigma_{DT}\rangle_0$	$k_B T_0$	ζ	ι	τ
8.745×10^{-16}	67.25	-5.1504×10^{-1}	2.5182×10^{-2}	-3.2849×10^{-3}

Table 2: Parameters being used for functional approximation of fusion cross section.

$\langle v\sigma_{DT}\rangle(k_B T)$ is shown in figure (7). Other functional approximations and approximations for cross sections of following fusion reactions are presented in [Atzeni04].

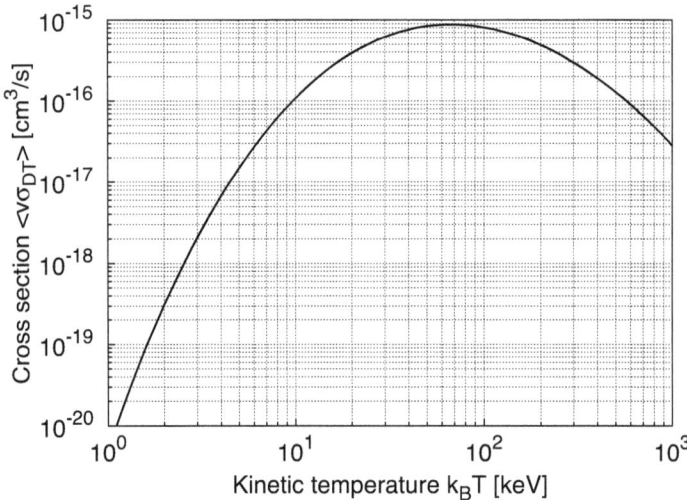

Figure 7: Cross section of the D-T reaction. The cross section as function depending on the kinetic temperature has been approximated by data given in [Duderstadt82]. The cross section has a maximum at around 70 keV.

At conditions like in the current physical model it is to be expected, that the thermonuclear burn-up will become very high. This has to be keep in mind while evaluating (3.5.5). Based on an idea of Hafner [Hafner09] a relation between the thermonuclear burn-up and fusion rate is derived. That concept is presented in details. To simplify matter equal particle densities for deuterium and tritium – $n_D = n_T = \tilde{n}$ have been assumed. In that way (3.5.5) reads

3.5 External Energy Contributions

$$R(r,t) = \tilde{n}^2(r,t) \langle v\sigma_{AB} \rangle (r,t). \tag{3.5.7}$$

$\tilde{n}$ has the meaning of a density of mass averaged fusion particles being available for the current fusion process. Consider the fusion rate $F(r,t)$ in a small comoving volume $dV(r,t)$

$$F(r,t) = \frac{d}{dt}dN_{fus}(r,t) = R(r,t)dV(r,t). \tag{3.5.8}$$

$dN_{fus}(r,t)$ being the number of fusions in $dV(r,t)$ occurred since a time t_0

$$dN_{fus}(r,t) = \int_{t_0}^{t} dt' F(r,t'). \tag{3.5.9}$$

Insertion of (3.5.7) in (3.5.8) gives

$$\frac{d}{dt}dN_{fus}(r,t) = \left(\frac{d\tilde{N}(r,t)}{dV(r,t)}\right)^2 \langle v\sigma_{DT} \rangle (r,t) dV(r,t). \tag{3.5.10}$$

Let a be the initial position of the small D-T plasma volume at some initial time t_0 and $u(r,t)$ the collective flow velocity. Neglecting the loss of mass due to fusion neutron productions a is called a Lagrangian coordinate. a and r are connected by

$$r(a,t) = a + \int_{t_0}^{t} dt' u(a,t'). \tag{3.5.11}$$

The periodic change dN_{fus} of fusion reactions in $dm(a)$ is in balance with the periodic change $d\tilde{N}$ of remaining D-T plasma pairs. Hence,

$$0 = \frac{dN(a,t)}{dt} = \frac{dN(a,t)_{fus}}{dt} + \frac{d\tilde{N}(a,t)}{dt} \tag{3.5.12}$$

$$\frac{d}{dt}dN_{fus}(a,t) = -\frac{d}{dt}d\tilde{N}(a,t). \tag{3.5.13}$$

Using the relation $dm(a) = \rho(a,t)dV(a,t)$ one finds[3]

$$-\frac{d}{dt}\frac{d\tilde{N}(a,t)}{dm(a)} = P(a,t), \tag{3.5.14}$$

where

---

[3] The mass of a comoving small volume of a fluid is constant in a Lagrangian coordinate system.

3 Physical model

$$P(a,t) = \left(\frac{d\tilde{N}(a,t)}{dm(a)}\right)^2 \rho(a,t) \langle v\sigma_{DT}\rangle(a,t) \tag{3.5.15}$$

being the fusion rate per mass. Solving the differential equation (3.5.14) and ordering terms yields

$$\frac{d\tilde{N}(a,t)}{dm(a)} = \frac{dN(a)}{dm(a)} \left[1 + \frac{dN(a)}{dm(a)} \int_{t_0}^{t} dt' \rho(a,t') \langle v\sigma_{DT}\rangle(a,t')\right]^{-1}. \tag{3.5.16}$$

$dN(a) = d\tilde{N}(a,t_0)$ is the number of initial D-T plasma pairs at some initial time t_0. The thermonuclear burn-up is defined by

$$B(a,t) = 1 - \tilde{B}(a,t) = 1 - \frac{d\tilde{N}(a,t)}{dN(a)}. \tag{3.5.17}$$

Hence, (3.5.16) becomes

$$\tilde{B}(a,t) = \left[1 + \frac{1}{m_{DT}} \int_{t_0}^{t} dt' \rho(a,t') \langle v\sigma_{DT}\rangle(a,t')\right]^{-1}. \tag{3.5.18}$$

Using $dm(a) = dN(a)m_{DT}$ the fusion rate per mass is associated by $\tilde{B}(a,t)$ via

$$P(a,t) = \left(\frac{\tilde{B}(a,t)}{m_{DT}}\right)^2 \rho(a,t) \langle v\sigma_{DT}\rangle(a,t). \tag{3.5.19}$$

If the thermonuclear burn up is neglectable $\tilde{B}(a,t)$ is constant and can be set to 1. Thus (3.5.19) reads[4]

$$P(a,t) = \frac{\rho(a,t)}{m_{DT}^2} \langle v\sigma_{DT}\rangle(a,t). \tag{3.5.20}$$

Within the program system relation $\tilde{B}$ appearing in (3.5.18) is solved by a recurrence relation and inserted into (3.5.19).

---

[4] (3.5.20) can also be derived directly from (3.5.5) by neglecting the changes in the number of particles of species a and b.

4 Ionisation energies

For the calculation of the bound-bound and bound-free cross section in section (6) one needs all bound energy levels of the accordant element. Formally, all energy levels can be obtained by solving the Schrödinger equation. An analytic solution of this equation is known for hydrogen only. However, a numerical solution method for high Z elements is implemented by [Blenski88] for example. The presented method is used for calculations of radiation opacities for high Z elements [Blenski90].

4.1 Ionisation energies for hydrogen and helium

The ionisation energy of hydrogen are obtained analytically by the equation of Schrödinger. Its value reads I_1^H=13.6 eV. The ionisation energies of helium can be found in data tables [Allen64]. Therein one finds $I_1^{He} = 24.59$ eV and $I_2^{He} = 54.4$ eV.

4.2 The model of Slater and its application on uranium and plutonium

Theoretically, the ionisation potentials of the elements are given by the solution of the equation of Schrödinger. Due to the complex interactions of the electrons in the atomic shells, an analytical solution is not known till now. One can find the ionisation energy for the outer electron in literature. For uranium one has $I_1^U = 6.08$ eV and for Plutonium $I_1^U = 5.8$ eV [Allen64]. The binding energy of the electron next to the core is obtained approximately by a hydrogenic model

$$I_Z = Z^2 \frac{m_e e^2}{2\hbar^2} = I_1^H \cdot Z^2. \quad (4.2.1)$$

Z is the nuclear charge. In that way one calculates $I_{92}^U = 115110$ eV and $I_{94}^{Pu} = 120170$ eV. It has been assumed very roughly $I_{91}^U \approx I_{91}^{Pa} \approx 112621$ eV and $I_{93}^{Pu} \approx I_{93}^{Np} \approx 117626$ eV. Between $r = 2$ and $r = 92$ the potentials of plutonium have been obtained by the model of Slater. In case of uranium the Slater model is applied between $r = 15$ and $r = 90$. Based on experimental data Slater [Slater30] has developed a phenomenological atomic model calculating the ionisation energies of an atom. Furthermore an effective quantum number, which results from a screening of the nuclear charge by electrons, is defined in table (3). Slater [Slater30] published the following algorithm to obtain the ionisation energies of all levels:

4 Ionisation energies

n	1	2	3	4	5	6	7
n*	1	2	3	3.7	4	4.2	4.3

Table 3: Comparison of the real principal quantum number n and the effective quantum number n^* in the model of Slater [Slater30]. Slater published values for n^* up to $n = 6$. An additional effective quantum number for heavy ions with nuclear charge larger than $Z = 92$ is needed. For $n = 7$ the effective quantum number $n^* = 4.3$ is roughly estimated by the extrapolation $n^*(n) = 0.566007 \, (488.497n - 1265.07)$.

1. The atomic configuration of an element is splitted into groups. Given the main quantum number n the s and p electrons belong to one groups and the d and f electrons form a group separately. Taking the electron configuration following from the orbital model (6) the group configuration for uranium and plutonium are given in tables (4) and (5). The shells are considered to be arranged from inside out in the order named.

2. The shielding constant s is formed, for any group of electrons, from the following contributions:

   - No contributions to s from any shell outside the one considered.
   - An amount of 0.35 from each other electron in the group is considered (if the group is the 1s group, use an amount of 0.30 instead.)
   - If the shell considered is a s or p shell, an amount of 0.85 from each electron with total quantum number less by one, and an amount 1.0 from each electron still further in. If the shell is a d or f shell, an amount 1.0 from every electron inside it.

Shell	s,p	d	f
1	2	0	0
2	8	0	0
3	8	10	0
4	8	10	14
5	8	10	3
6	8	1	0
7	2	0	0

Table 4: Electron configuration of uranium by the model of Slater.

Shell	s,p	d	f
1	2	0	0
2	8	0	0
3	8	10	0
4	8	10	14
5	8	10	6
6	8	0	0
7	2	0	0

Table 5: Electron configuration of plutonium by the model of Slater.

4.2 THE MODEL OF SLATER AND ITS APPLICATION ON URANIUM AND PLUTONIUM

E	1s	2s	2p	3s	3p	3d	4s	4p	4d	4f	5s	5p	5d	5f	6s	6p	6d	6f	7s	7p
U	2	2	6	2	6	10	2	6	10	14	2	6	10	3	2	6	1	0	2	0
Pu	2	2	6	2	6	10	2	6	10	14	2	6	10	6	2	6	0	0	2	0

Table 6: Electron configuration by the orbital model for uranium and plutonium. No interchange between shells has been applied. Normally, this has to be done for 4f/5d. The energy of shell 4f is larger than that one of 5d.

For our purpose the effective nuclear charge $Z_{eff} = Z - s$ for uranium $Z = 92$ is calculated by

1s $\quad Z_{eff,1} = 92 - 0.30 = 91.7$

2s,p $\quad Z_{eff,2} = 92 - 1 \times 0.35 - 8 \times 0.85 - 2 \times 1.00 = 82.85$

...

3d $\quad Z_{eff,4} = 92 - 9 \times 0.35 - 18 \times 1.00 = 70.85$

....

The total energy of bound electrons of an atom or ion is than given by

$$E_{tot} = \left[-\sum_{i=1}^{G} N_{e,i} \left(\frac{Z_{eff,i}}{n_{eff,i}} \right)^2 \right] \times 13.6 \text{ eV}, \qquad (4.2.2)$$

G is the number of groups, $N_{e,i}$ is the number of electrons belonging to group i and $Z_{eff,i}$ is the effective nuclear charge of group i. The ionisation energies are accomplished by taking the difference between two different ionisation levels. This is a straight forward but complex task for heavy ions. For that reason and to get confidence with this model the calculations have been done by a self written program. The results for uranium and plutonium are presented in tables (7) and (8).

4 IONISATION ENERGIES

r	I_r [eV]	r	I_r [eV]	r	I_r [eV]	r	I_r [eV]
1	6*	24	489	47	2443	70	7659
2	12*	25	859	48	2515	71	7812
3	26*	26	899	49	2588	72	7966
4	41*	27	939	50	2661	73	8121
5	58*	28	980	51	2735	74	8277
6	79*	29	1022	52	2810	75	9257
7	104*	30	1064	53	2885	76	9426
8	121*	31	1107	54	2961	77	9596
9	137*	32	1151	55	3038	78	9767
10	162*	33	1330	56	3116	79	9940
11	183*	34	1385	57	3788	80	10113
12	203*	35	1441	58	3876	81	10288
13	224*	36	1498	59	3965	82	10464
14	244*	37	1555	60	4055	83	24782
15	257	38	1613	61	4145	84	25196
16	280	39	1671	62	4236	85	25613
17	304	40	1731	63	4328	86	26032
18	328	41	1791	64	4421	87	26454
19	354	42	1852	65	6911	88	26878
20	380	43	1913	66	7059	89	27304
21	406	44	1976	67	7207	90	27734
22	433	45	2039	68	7357	91	130000†
23	461	46	2102	69	7507	92	132400‡

Table 7: Ionisation energies of uranium by using the atomic model of Slater [Slater30]. *Published values [Pritzker71], †Rough estimation by 90 times ionised Proactinium, ‡Hydrogen-like model (4.2.1).

4.2 The model of Slater and its application on uranium and plutonium

r	I_r [eV]	r	I_r [eV]	r	I_r [eV]	r	I_r [eV]
1	6*	25	544	49	2645	73	8253
2	7	26	574	50	2719	74	8411
3	53	27	971	51	2795	75	8570
4	64	28	1013	52	2871	76	8731
5	75	29	1056	53	2948	77	9736
6	88	30	1099	54	3025	78	9910
7	100	31	1144	55	3103	79	10084
8	114	32	1188	56	3182	80	10260
9	128	33	1234	57	3262	81	10436
10	142	34	1280	58	3343	82	10614
11	129	35	1481	59	4038	83	10793
12	146	36	1539	60	4129	84	10973
13	163	37	1597	61	4220	85	25957
14	180	38	1657	62	4313	86	26381
15	198	39	1717	63	4406	87	26807
16	217	40	1777	64	4500	88	27236
17	320	41	1839	65	4595	89	27667
18	346	42	1901	66	4690	90	28101
19	372	43	1964	67	7327	91	28537
20	399	44	2027	68	7478	92	28976
21	427	45	2092	69	7631	93	118681[†]
22	455	46	2157	70	7785	94	120213[‡]
23	484	47	2223	71	7940		
24	513	48	2289	72	8096		

Table 8: Ionisation energies of plutonium by using the atomic model of Slater [Slater30]. *Published values [Allen64], [†]Rough estimation by 92 times ionised Neptunium, [‡]Hydrogen-like model (4.2.1).

4 IONISATION ENERGIES

4.3 Equilibrium of ionisation - the equation of Saha

Assuming that the plasma is in local thermal equilibrium the averaged ionisation and the number of free electrons is given by solving the equation of Saha. Taking the plasma to be nonrelativistic and nondegenerated the equation of Saha is given by [Landau87]

$$\frac{n_r n_e}{n_{r-1}} = \frac{u_r}{u_{r-1}} \frac{2(2\pi m_e k_B T)^{3/2}}{h^3} \exp\left(\frac{-I_r}{k_B T}\right) \qquad r = 1, 2, \ldots, Z \qquad (4.3.1)$$

I_r is the ionisation energy, which is required to remove an electron from an atom, which is r-1-times ionised already. m_e is the electron mass, T the temperature, h the Planck constant, k_B the Boltzmann constant, n_r the number density of r-times ionised atoms and n_{r-1} the number density of r-1-times ionised atoms. u_r are the functions of states, which are formally given by

$$u_r = \sum_{k=1}^{\infty} g_{rk} \exp\left(-w_{rk}/k_B T\right). \qquad (4.3.2)$$

g_{rk} is the degenerate factor and w_{rk} is the excitation energy of the kth atomic level. Saha's equation is being normalised to

$$n_{tot} = \sum_{r=0}^{Z} n_r, \qquad (4.3.3)$$

where n_{tot} is the total particle density of atoms/ions. The particle density for free electrons is obtained by

$$n_e = \sum_{r=1}^{Z} r n_r. \qquad (4.3.4)$$

n_r is the particle number density in cm^{-3} of the ionisation level r. The conditions (4.3.3) and (4.3.4) describe the conservation of particles and charge of the plasma. Defining the average ionisation stage $\bar{r}$ by

$$\bar{r} = \frac{n_e}{n_{tot}}, \qquad (4.3.5)$$

the participation probability n_r^* of the r-times ionised atoms to the ionisation equilibrium by

$$n_r^* = \frac{n_r}{n_{tot}}. \qquad (4.3.6)$$

4.3 EQUILIBRIUM OF IONISATION - THE EQUATION OF SAHA

and

$$
\begin{aligned}
K_r &= \frac{u_r}{u_{r-1}} \frac{2(2\pi m_e k_B T)^{3/2}}{n_{tot} h^3} \exp(-I_r/k_B T) \\
&= 6.04 \times 10^{21} \frac{u_r}{u_{r-1}} \frac{T^{3/2}}{n_{tot}} \exp(-I_r/k_B T) \quad [\text{eV}^{-3/2}\text{cm}^{-3}],
\end{aligned}
\qquad (4.3.7)
$$

the system of equations (4.3.1) is rewritten to

$$\frac{n_r^*}{n_{r-1}^*} = \frac{K_r}{\bar{r}}. \qquad (4.3.8)$$

One finds a closed solution of the equation of Saha for a hydrogen plasma. Starting from (4.3.3) one has

$$1 = n_0^* + n_1^*. \qquad (4.3.9)$$

Taking $r = 1$ in (4.3.8) and using the above result one finds

$$1 = n_0^* \left(1 + \frac{K_1}{\bar{r}}\right). \qquad (4.3.10)$$

For D-T $Z = 1$ the averaged ionisation $\bar{r}$ is given by $\bar{r} = n_1^*$. Together with (4.3.8) one finds

$$\frac{\bar{r}^2}{K_1} = n_0^*. \qquad (4.3.11)$$

Eliminating n_0^* leads to the quadratic equation

$$\bar{r}^2 + K_1 \bar{r} - K_1 = 0. \qquad (4.3.12)$$

The solution is

$$\bar{r} = -\frac{K_1}{2} + \sqrt{\frac{K_1^2}{4} + K_1} = \frac{K_1}{2}\left(\sqrt{1 + \frac{4}{K_1}} - 1\right). \qquad (4.3.13)$$

u_0 and u_1 appearing in K_1 are given by [Zel'dovich66]

$$u_0 = \sum_{m=0}^{m^*} 2m^2 \exp\left(\frac{-I_H}{k_B T}\left(1 - \frac{1}{m^2}\right)\right), \qquad u_1 = 1, \qquad (4.3.14)$$

where $I_H = 13.6$ eV. The summation in u_0 is truncated at a value m^*, when the radius of

4 IONISATION ENERGIES

the m-th Bohr orbit is larger then the intermediate distance between hydrogen atoms/ions. Let $1/n_H$ be the volume of a hydrogen atom approximately. n_H is the number density of hydrogen atoms. Hence, the atomic radius is calculated to be $r \approx 0.62 \times n_H^{-1/3}$. The radius of the m-th Bohr orbit is given by [Zel'dovich66] $r_B = a_0\, m^{*2}$ where a_0 is the Bohr radius. The truncation condition for m^* reads $m^* = [N]$. $[N]$ is the largest integer value smaller than $1.083 \times 10^4\, n_H^{-1/6}$. At low particle densities ($n \leq 10^{19}$ cm^{-3}) the D-T plasma is immediately ionised. With increasing density the radius of the outermost electron orbit becomes smaller. A decreasing outermost electron orbit is related to a decreasing averaged ionisation stage $\bar{r}$. The averaged degree of ionisation for D-T depending from temperature and density is shown in figure (8).

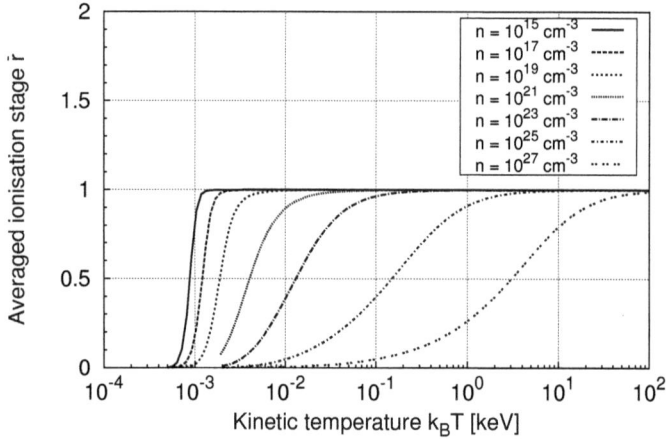

Figure 8: Averaged ionisation stage for a hydrogen plasma depending on temperature and density.

For elements with large nuclear charges the functions of state u_r are unknown. It is shown that in these cases the ratio $u_r/u_{r-1} \approx 1$ is an useful approximation [Pritzker71]. The leading 2 in (4.3.1) vanishes. The system of equations (4.3.1) describes a system of Z nonlinear equations. Following an idea of [Leuthäuser68] this system is transformed to an nonlinear equation of degree Z. Division of (4.3.4) by the total density n_{tot} leads together with (4.3.6) to

4.3 Equilibrium of Ionisation - the Equation of Saha

$$\bar{r} = \frac{n_e}{n_{tot}} = \sum_{r=1}^{Z} r \frac{n_r}{n_{tot}} = \sum_{r=1}^{Z} r n_r^*. \qquad (4.3.15)$$

Equation (4.3.3) is transformed to

$$1 = \sum_{r=0}^{Z} n_r^*. \qquad (4.3.16)$$

The equation of Saha reads

$$\frac{n_1^*}{n_0^*} = \frac{K_1}{\bar{r}}$$
$$\frac{n_2^*}{n_1^*} = \frac{K_2}{\bar{r}} \Rightarrow \frac{n_2^*}{n_0^*} = \frac{K_2 K_1}{\bar{r}^2}$$
$$\frac{n_3^*}{n_2^*} = \frac{K_3}{\bar{r}} \Rightarrow \frac{n_3^*}{n_0^*} = \frac{K_3 K_2 K_1}{\bar{r}^3}$$
$$\vdots$$
$$\Rightarrow \frac{n_j^*}{n_0^*} = \frac{1}{\bar{r}^j} \prod_{i=1}^{j} K_i \qquad j = 1, \ldots, Z. \qquad (4.3.17)$$

Inserting the latter result in (4.3.15) yields

$$\bar{r} = n_0^* \sum_{j=1}^{Z} \frac{j}{\bar{r}^j} K_1 \cdots K_j = n_0^* \sum_{j=1}^{Z} \frac{j}{\bar{r}^j} \prod_{i=1}^{j} K_i. \qquad (4.3.18)$$

Using (4.3.16) results in

$$1 = n_0^* \left(1 + \sum_{j=1}^{Z} \frac{1}{\bar{r}^j} K_1 \cdots K_j\right) = n_0^* \left(1 + \sum_{j=1}^{Z} \frac{1}{\bar{r}^j} \prod_{i=1}^{j} K_i\right). \qquad (4.3.19)$$

Finally, merging the latter equations gives the final result [Leuthäuser68]

$$\bar{r} + \sum_{j=1}^{Z} \frac{1}{\bar{r}^j} (\bar{r} - j) \prod_{i=1}^{j} K_i = 0. \qquad (4.3.20)$$

The advantage of (4.3.20) is that this equation is independent of the choice of n_0^*, whereas solving a system of equations as given by (4.3.1) is not. However, (4.3.20) is numerically complicated to solve. In the particular case of heavy ions small $\bar{r}$ and the product of K_i at large j lead to numbers, which cannot be represented by a computer. A new strategy has been developed by calculating the mantissa and exponents of all terms in (4.3.20) sep-

4 IONISATION ENERGIES

arately. After the separation procedure the root of (4.3.20) is calculated iteratively by a bisection algorithm.

It should be emphasised, that solving equation (4.3.20) is one of the main aspects within this thesis. All macroscopic radiative coefficients depend on knowledge of the free electron density, which depends on $\bar{r}$. Further, the participation probability n_r^* of r-times ionised atoms is required for the determination of the bound-free coefficient.

The electrostatic potential of an ion surrounded by a plasma is influenced by its bound electrons, free electrons, free ions and by bound electrons of other particles. An ion influenced by these interactions requires less energy to be ionised. Using the result of [Cox68b] one finds

$$D = 1.1605751 \times 10^{-10} \frac{n_e}{n_{tot}^{2/3}} \text{ [keV]}. \quad (4.3.21)$$

For studying the influence of depression energies K_i is extended to

$$K_i^D = 2 \frac{(2\pi m_e k_B T)^{\frac{3}{2}}}{h^3 n_{tot}} \exp\left(-(I_i - D)/k_B T\right) \quad [\text{eV}^{-3/2}\text{cm}^{-3}]. \quad (4.3.22)$$

The influence of depression energy on the averaged ionisation stage of uranium and plutonium is shown in figures (9) and (11). The influence is of small effect and is neglected within this thesis. The participation probabilities of r-times ionised uranium atoms at normal particle density in the ionisation equilibrium is shown in figure (10).

Other models of lowering of the ionisation energy for a plasma in thermodynamic equilibrium are set up in [Zimmermann79, Ecker63, Kunc92]. Other effects, which become the focus of attention at high temperatures ($k_B T \geq 10$ keV), like the pair production will be neglected within all investigations.

4.4 Advanced methods

The scheme of Slater is only suitable for a rough estimation of ionisation energies. In the last years accurate numerical methods have been developed to solve the equation of Schrödinger. Gonzalez et al. [Gonzalez97] present a first step method to initialise the general algorithm of Numerov. That algorithm solves accurately second order differential equations with no first derivative. Such an equation is the time-independent Schrödinger equation. Blenski et al. [Blenski90] apply the scheme of Numerov to evaluate the electron energy levels and wave functions. The electron energy levels as well as the wave functions are necessary for obtaining the bound-bound, bound-free and free-free contributions to the radiation cross sections. Their method has been described in detail in [Blenski88]. However, these methods are more accurate, but are more expensive in programming and numerical evaluation. The methods used within this thesis are restricted to classical approximations to ionisation energies and radiation cross sections.

4 IONISATION ENERGIES

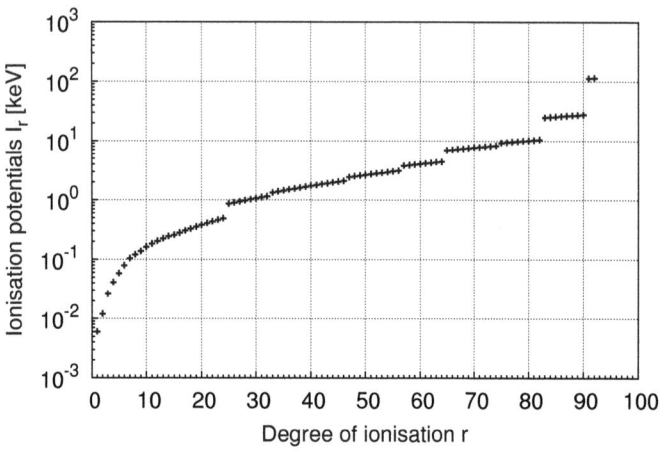

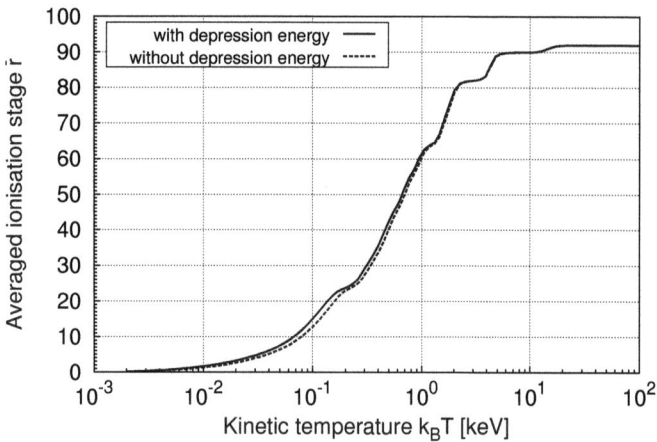

Figure 9: Comparison of the different ionisation stages of uranium at normal particle density $n = 4.78 \times 10^{22}$ cm^{-3} depending on temperature and depression energies. Recognising the small influence of the depression energy and the roughly estimated ionisation energies the depression energy term will be ignored within the thesis.

4.4 ADVANCED METHODS

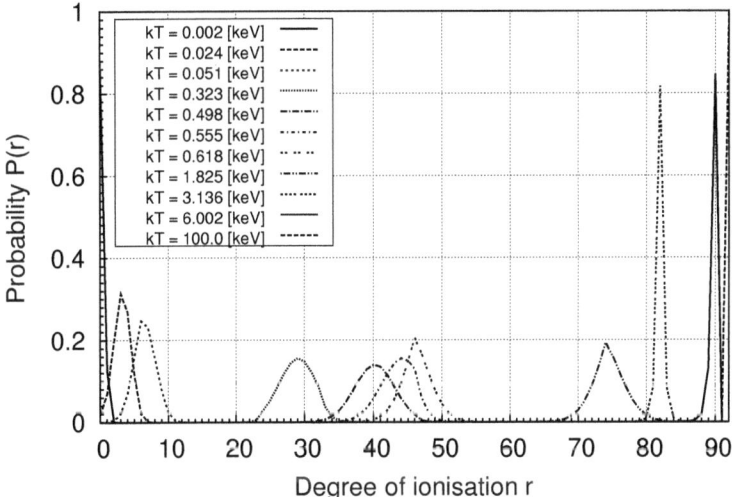

Figure 10: Examples of participation probabilities of r-times ionised uranium atoms at normal particle density in the ionisation equilibrium. Regarding the temperature sequence $k_BT = 0.498, k_BT = 0.555, k_BT = 0.618$ and comparing to figure (9) one discovers, that the probability becomes sharper and higher, when the temperature is close to the ionisation potential of a closure of a shell.

4 IONISATION ENERGIES

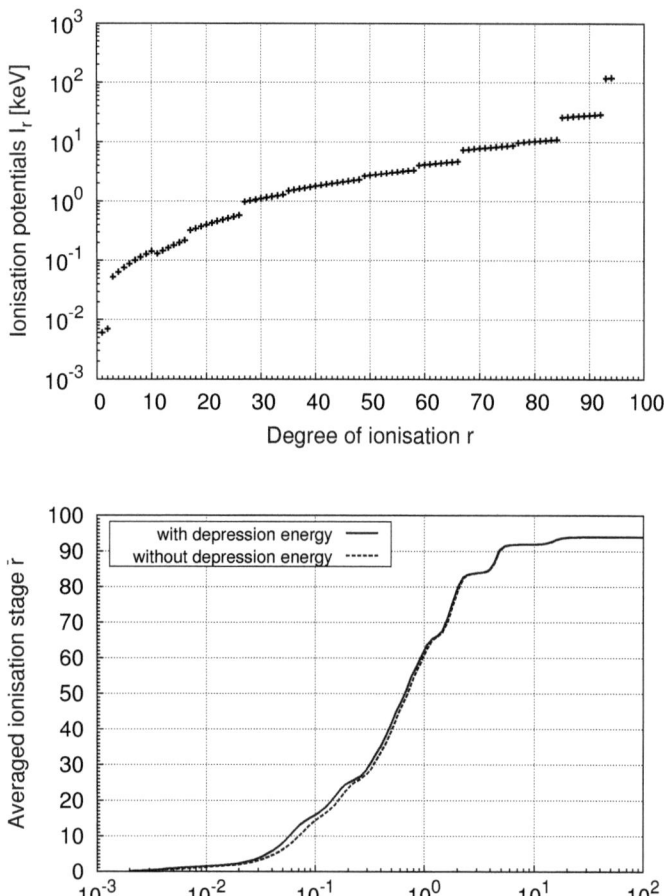

Figure 11: Comparison of the different ionisation stages of plutonium at normal particle density n = 4.87×10^{22} cm^{-3} depending on temperature and depression energies. Recognising the small influence of the depression energy and the roughly estimated ionisation energies the depression energy term will be ignored within the thesis. The jump at $r = 11$ to lower ionisation potentials is unphysical. This has to be understood as a problem by the enhancement of the Slater model of an additional effective quantum number. However, the error is of small effect.

5 Radiation cross sections

At high temperatures the encapsulated material and radiation are affected by each other. Solving the coupled radiation transport and hydrodynamic problem requires the knowledge of the radiation cross sections on which the equation of radiative transfer depends. That cross sections are specific for each constituent of a plasma and include absorption, emission and scattering contributions. It is necessary to distinguish between the process of absorption and scattering. A process is called *scattering* when a photon interacts with a scattering centre (atom, ion, electron) and emerges from the interaction in a new direction with no altered energy (Thomson scattering) or a slightly altered energy (Compton scattering). A process is called absorption, when a photon is destroyed by conversion of its energy (wholly or partly) into thermal energy of the plasma [Mihalas78]. Sometimes those absorption processes are called *true absorption*. Scattering processes depend mainly upon the radiation field and are only weakly coupled to the thermodynamic properties of the plasma. Absorption processes convert the photon energy directly into thermal energy of the gas [Mihalas78].

For astrophysical investigations elements with nuclear charge smaller than 26 are important. In ICF calculations opacities of high Z elements like gold or lead, which act as pellet tamper, are required. Finally, for uncontrolled thermonuclear explosions the knowledge of opacities for uranium and plutonium are essential. Radiation properties for high Z elements are rare in published literature [Blenski90]. Pritzker et al. [Pritzker75, Pritzker76] approximate the opacities for uranium based on a hydrogenic atomic model, but they ignore the bound-bound transitions, which have been recognised as giving dominant contributions to opacities for high Z elements [Blenski90].

Early opacity codes, e.g. [Tsakiris87], take the *average ion model* [Mayer47] into account. That model is an enhancement based on the idea of Slater [Slater30] by considering an individual electron in a complex atom subjected to an effective nuclear charge.

A modern method for determining opacities is the UTA – unresolved transition array – method [Duffy91]. Transition array means all the lines between specified electronic configurations. By the UTA method each transition array is treated as a single, broad, unresolved spectral feature. This is often approximated by a Gaussian distribution. Especially, this method is particularly suitable for calculations of the bound-bound atomic transitions. Zeqing et al. [Zeqing06] calculate opacity data for nonlocal thermal equilibrium plasma

5   RADIATION CROSS SECTIONS

by solving the rate equation. In their publication the UTA approximation has been used while evaluating absorption spectra.

In the present consideration classical methods have been applied to obtain absorption and scattering cross sections [Zel'dovich66, Pomraning73]. Due to the assumption of a local thermal equilibrium the emission and absorption coefficient are connected by the rule of Kirchhoff [Kourganoff63].

The mechanism of absorption of radiation by matter are the bound-bound absorption, bound-free absorption (photo effect) and free-free absorption (bremsstrahlung). See figure (12).

Often one introduces frequency integrated cross sections. In that case radiation cross sections depend mainly on temperature only. That method is called a one-group model or grey model and often simplifies matter. A procedure for a simplified handling of the frequency dependent variables in the equation of radiative transfer is the multigroup method. Refer to section 7. For multigroup models frequency group averaged cross sections are required. For that the whole frequency spectrum is splitted into several groups. The cross sections belonging to one frequency group are grouped together and averaged by an appropriate method over the chosen frequency interval. In a one-group approximation the averaged method of Rosseland and Planck has been established.

Within this section the influence of the different contributions to the absorption cross section have been studied. Concrete formulas for free-free absorption (bremsstrahlung), the combined free-free + bound-free absorption (bremsstrahlung + photo effect) and scattering cross section are given. For the purpose of radiation heat conduction frequency averaged mean free paths are derived.

5.1   Bremsstrahlung/Inverse bremsstrahlung

The emission of a photon due to the slowing down of an electron in the electric fields of ions is called *bremsstrahlung*. The inverse process the absorption of a photon by an free electron within the field of an ion is called *inverse bremsstrahlung*. Both processes are referred as *free-free* because the electron is unbound before and after absorption of a photon [Mihalas78]. The classical frequency dependent free-free transition coefficient is presented by Kramer's formula, whereupon the macroscopic coefficient is given by [Zel'dovich66]

5.1 BREMSSTRAHLUNG/INVERSE BREMSSTRAHLUNG

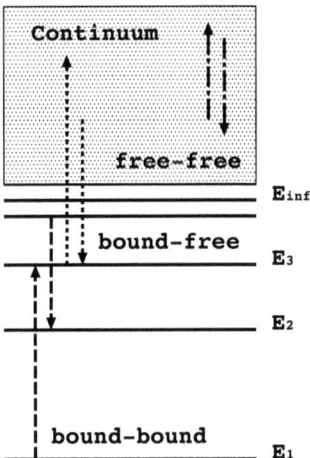

Figure 12: Different transition contribution to the absorption cross section. E_1, E_2, ... E_{inf} show the discretise energy levels.

$$\Sigma_{ff}(\nu) = \frac{4}{3}\left(\frac{2\pi}{3m_e\theta}\right)^{1/2}\frac{Z^2 e^6}{hcm_e\nu^3}n_i n_e \tag{5.1.1}$$

$$\Rightarrow \Sigma_{ff}(u) = 0.767 \times 10^{-47}\frac{Zn_e^2}{u^3\theta^{7/2}}\quad\left[\text{cm}^{-1}\right] \tag{5.1.2}$$

$$\Rightarrow \Sigma_{ff}(u) \approx 0.767 \times 10^{-47}\frac{\bar{r}^3 n_i^2}{u^3\theta^{7/2}}\quad\left[\text{cm}^{-1}\right], \tag{5.1.3}$$

where $\theta = k_B T$, $u = h\nu/\theta$, n_i is the ion particle density and $\bar{r}$ is the averaged ionisation stage. The electron density n_e and $\bar{r}$ are determined from the Saha equation (4.3.1). In the current ion sphere model the ion density n_i and the initial particle density n_{tot} are equal. Initial particle density mean the number of all atoms per volume.

To correct the semi-classical result for quantumechanical influences an additional factor, the Gaunt factor is introduced in (5.1.2) [Pomraning73]. Considerable effort have been made by Karzas and Latter [Karzas61], whose computed the free-free, bound-free and bound-bound Gaunt factors for electrons in a pure Coulomb field. The approximate Gaunt factor g_{ff} for free-free transitions has been taken into account

5 Radiation cross sections

$$g_{ff} = 1.0 + 0.1728 \left(\frac{h\nu}{13.6\bar{r}^2}\right)^{1/3} \left(1.0 + 2.0\frac{\theta}{h\nu}\right). \tag{5.1.4}$$

The free-free absorption coefficient for isotopes of hydrogen, where the ionic charge is set to one, is given by (5.1.2)

$$\Sigma_{ff} = 0.767 \times 10^{-47} \frac{n_e^2}{u^3 \theta^{7/2}} \quad [\text{cm}^{-1}] \tag{5.1.5}$$

Khalfaoui et at. [Khalfaoui97] consider a model, where nonideal effects of conduction are taken into account. Such a nonideal effect is the Fermi degeneracy of plasma at low temperatures and high densities. The parameter of degeneracy is defined by $\bar{\theta} = \theta/E_F$, where $E_F = (\hbar/2m_e) \times (3\pi^2 n_e)^{2/3}$ is the Fermi energy. For $\bar{\theta} \ll 1$ the system is complete and for $\bar{\theta} \approx 1$ intermediate degenerated. For degenerate plasma, the absorption coefficient must be averaged over a Fermi-Dirac distribution. Following Cox et al. [Cox68a] the multiplicative factor in the free-free absorption coefficient is

$$\Phi = \frac{\sqrt{\pi}}{2} \frac{\log\left(\frac{1+\exp(\alpha)}{1+\exp(\alpha - h\nu/\theta)}\right)}{F_{0.5}(-\alpha)(1 - \exp(-h\nu/\theta))} \tag{5.1.6}$$

where $\alpha = \mu/\theta$ and $F_{0.5}(-\alpha)$ is the Fermi-Dirac integral. μ is the chemical potential. Due to the surge in temperature in the presented physical model the degeneracy is assumed to be of small influence. Hence, the influence of degeneracy of electrons in the absorption coefficient will be neglected. Studying this effect might be a task for the future.

5.2 Total absorption coefficient

At high temperatures heavy materials are not completely ionised. In a wide range of temperature one has to consider bound-free contributions to the free-free coefficient additionally. A process is called bound-free absorption or photoionisation when a photon is absorbed by an atom and ionises a bound electron allowing it to escape with finite kinetic energy into the continuum [Mihalas78]. The frequency dependent total absorption coefficient is estimated by [Zel'dovich66]

$$\Sigma_a(\nu) = \sum_{m=0}^{Z-1} \Sigma_a^m(\nu), \tag{5.2.1}$$

where the parts $\Sigma_a^m(\nu)$ are given by

$$\Sigma_a^m(\nu) = \begin{cases} \dfrac{16\pi^2}{3\sqrt{3}} \dfrac{e^6 Z^2}{hc} n_m \dfrac{\theta}{(h\nu)^3} \exp\left(\dfrac{h\nu}{\theta} - \dfrac{I_{m+1}}{\theta}\right) & : \dfrac{h\nu}{\theta} \leq \dfrac{I_{m+1}}{\theta} \\ \dfrac{16\pi^2}{3\sqrt{3}} \dfrac{e^6 Z^2}{hc} n_m \dfrac{2}{(h\nu)^3} I_{m+1} & : \dfrac{h\nu}{\theta} > \dfrac{I_{m+1}}{\theta} \end{cases} \quad (5.2.2)$$

n_m is the number density of the m-times ionised atoms. For numerical purpose the bound-free absorption contribution is transformed to

$$\Sigma_a(u) = 0.96 \times 10^{-7} \dfrac{n_{tot}}{T^2} \sum_{m=0}^{Z-1} n_m^*(m+1)^2 F_m(u) \quad (5.2.3)$$

where $n_m^* = n_m/n_{tot}$ and $u = h\nu/\theta$. $Z = m - 1$ is the residual charge of the m-times ionised atoms and $F_m(u)$ given by

$$F_m(u) = \begin{cases} \dfrac{1}{u^3} \exp\left(u - \dfrac{I_{m+1}}{\theta}\right) & : u \leq \dfrac{I_{m+1}}{\theta} \\ \dfrac{2}{u^3} \dfrac{I_{m+1}}{\theta} & : u > \dfrac{I_{m+1}}{\theta}. \end{cases} \quad (5.2.4)$$

The temperature T is measured in Kelvin. (5.2.2) is known as Kramer-Unsöld formula. Pritzker et al. [Pritzker75] have mentioned a macroscopic bound-free absorption coefficient including first order corrections to excited atomic states.

5.3 Effects of scattering of photons on electrons

The scattering of low energy photons on free electrons at rest are described by the Thomson coefficient [Rutten03]

$$\Sigma_{Th} = \sigma_{Th} n_e = \dfrac{8\pi e^4}{3 m_e^2 c^2} n_e = 0.665 \times 10^{-24} n_e \quad [\text{cm}^{-1}] \quad (5.3.1)$$

where σ_{Th} is the frequency independent microscopic Thomson coefficient and n_e is the electron density. Obviously, no shift in frequency of photons occurs. That means the photon is scattered elastically on the electron. No energy transfer to the electron occurs. In case of a photon energy in the range of the electron plasma temperature ($h\nu \approx \theta$) one has to use the frequency dependent Klein-Nishina formula [Heitler57]. The integration over all final frequencies ν' and scattered directions $\mathbf{\Omega}'$ gives by help of the definition $\gamma = h\nu/m_e c^2$

5 Radiation cross sections

$$\Sigma_s(\gamma) = \frac{3}{4}\Sigma_{Th}\left[\left(\frac{1+\gamma}{\gamma^3}\right)\left(\frac{2\gamma(1+\gamma)}{1+2\gamma} - \log(1+2\gamma)\right)\right.$$
$$\left. + \frac{1}{2\gamma}\log(1+2\gamma) - \frac{1+3\gamma}{(1+2\gamma)^2}\right]. \qquad (5.3.2)$$

5.4 Mean free path methods

The mean free path of a photon is a measure at which distance, on the average, a photon undergoes a collision.[5] The mean free path λ is related to the cross section Σ by $\lambda = \Sigma^{-1}$. Mean free paths are of great interest in inertial confinement fusion (ICF), astro- and plasma physics for developing approximative tools for investigation of radiative transfer in media. Depending from the material density and temperature a system can be transparent or opaque for radiation. In the first case the system is called optical thin in the other case optical thick. One can also say, that a system is optical thin, when the radiation mean free path is of order or larger than the dimension of the physical system. Such system are almost transparent for radiation. A system is optical thick, when the radiation mean free path is much smaller than the dimensions of the system. In an optical thin regime the frequency averaged mean free path is determined by the method of Planck, in an optical thick regime one has to use the method of Rosseland [Zel'dovich66]. The Rosseland averaged mean free path is used for an estimation of the radiation heat conductivity coefficient. See section (6.1).

5.4.1 Optical depth

A quantity, which is helpful to distinguish between optical thick and thin systems, is the optical thickness $\tau^{\ddagger}$

$$\frac{d\tau_\nu}{dr} = -\Sigma_a(\nu). \qquad (5.4.1)$$

A plasma is denoted as optical thin, when $|\tau| \ll 1$. In that case the absorption is small and the mean free path of photons is in the range of the dimension of the system. On the other hand a system is optical thick when the mean free paths are small. In this limit the

---
[5]That can be true absorption or scattering.
‡In grey systems τ is defined by
$$\frac{d\tau}{dr} = -\Sigma_a(T).$$

5.4 MEAN FREE PATH METHODS

photons are absorbed in the vicinity of their origin. If a plasma belongs to one of both categories the calculations of the mean free paths are given by the methods of Rosseland and Planck.

5.4.2 Rosseland and Planck limit

The Rosseland mean free path is defined as

$$\lambda_R = \frac{\int_0^\infty \frac{d\nu}{\Sigma_{tr}} \frac{\partial B_\nu}{\partial \theta}}{\int_0^\infty d\nu \frac{\partial B_\nu}{\partial \theta}} = \frac{15}{8\pi^4} \frac{c^2 h^3}{\theta^3} \int_0^\infty \frac{d\nu}{\Sigma_{tr}} \frac{\partial B_\nu}{\partial \theta}. \tag{5.4.2}$$

where Σ_{tr} is the transport cross section and B_ν is Planck's function (3.2.2). Recognising

$$\frac{\partial B}{\partial \theta} = \frac{2h\nu^3}{c^2} (\exp(h\nu/\theta) - 1)^{-2} \frac{h\nu}{\theta^2} \exp(h\nu/\theta),$$

inserting (3.2.2) and evaluating the resulting integrals one finds

$$\lambda_R = \frac{15}{8\pi^4} \frac{c^2 h^3}{\theta^3} \int_0^\infty \frac{d\nu}{\Sigma_{tr}} \frac{2h\nu^3}{c^2} (\exp(h\nu/\theta) - 1)^{-2} \frac{h\nu}{\theta^2} \exp(h\nu/\theta). \tag{5.4.3}$$

It is of practical interest to introduce the dimensionless variable $u = h\nu/\theta$. In that way (5.4.3) becomes

$$\lambda_R = \frac{15}{4\pi^4} \int_0^\infty \frac{du}{\Sigma_{tr}} u^4 \exp(-u) (1 - \exp(-u))^{-2} \tag{5.4.4}$$

In general (5.4.4) is solved numerically. The transport cross section Σ_{tr} includes contributions from absorption, total scattering and the first moment of the scattering coefficient [Pomraning73].

In optical thin regimes the transport cross section has to be averaged by the method of Planck. In that case the mean free path is given by

$$\lambda_P = \int_0^\infty d\nu\, B_\nu(\theta) \times \left(\int_0^\infty d\nu\, \Sigma_{tr} B_\nu(\theta) \right)^{-1} \tag{5.4.5}$$

Inserting Planck's function and evaluating the leftmost integral by help of (A.3.6) gives

$$\lambda_P = \frac{\pi^4}{15} \left(\int_0^\infty du\, \Sigma_{tr} u^3 \exp(-u) (1 - \exp(-u))^{-1} \right)^{-1}. \tag{5.4.6}$$

One obtains an analytical solution of λ_R and λ_P by taking only free-free transitions into

5 Radiation cross sections

account, that means neglecting scattering, bound-free contributions and the Gaunt factor. In that way using (5.1.2) in (5.4.4) yields

$$\lambda_R = \frac{15}{4\pi^4} \int_0^\infty \frac{du}{\Sigma_{ff}} \frac{u^4 \exp(-u)}{(1 - \exp(-u))^3}$$

$$\Rightarrow \lambda_R = \frac{15}{4\pi^4} 1.3038 \times 10^{47} \frac{\theta^{7/2}}{\bar{r}^3 n_i^2} \int_0^\infty du \frac{u^7 \exp(-u)}{(1 - \exp(-u))^3}. \qquad (5.4.7)$$

The Integral $\int_0^\infty du \frac{u^7 \exp(-u)}{(1 - \exp(-u))^3}$ is evaluated numerically and has a value of 5104.7. Hence,

$$\Sigma_R'^{-1} = \lambda_R = 2.56217 \times 10^{49} \frac{\theta^{7/2}}{\bar{r}^3 n_i^2} \quad [\text{cm}]. \qquad (5.4.8)$$

Similarly, the Planck averaged mean free path taking free-free transitions into account reads

$$\lambda_P = \frac{\pi^4}{15} \left(\int_0^\infty du\, \Sigma_{ff} (1 - \exp(-u))\, u^3 (\exp(u) - 1)^{-1} \right)^{-1}$$

$$\lambda_P = \frac{\pi^4}{15} 1.3038 \times 10^{47} \frac{\theta^{7/2}}{\bar{r}^3 n_i^2} \left(\int_0^\infty du\, \exp(-u) \right)^{-1}$$

$$\Sigma_P'^{-1} = \lambda_P = 8.46679 \times 10^{47} \frac{\theta^{7/2}}{\bar{r}^3 n_i^2} \quad [\text{cm}]. \qquad (5.4.9)$$

In figure (13) the Rosseland mean free path of photons in a plasma consisting of D-T and uranium ions depending on temperature and density is shown. Free-free transitions have been considered only. $n_{tot} = 4.78 \times 10^{22}$ cm^{-3} is the normal particle density of uranium. The Rosseland approximation becomes invalid in the limiting case where $\lambda_R \approx R$, where R is the outer radius of the present physical model. Depending on the special configuration R is in the range of several meters. Hence, in figure (13) the Rosseland averaged mean free path becomes invalid in regions with high temperatures and low densities. Therefore, the proposed Rosseland averaged free-free absorption coefficient (5.4.8) is valid only in regions with high particle densities and moderate temperatures. This is the case in stellar envelopes close to the centre of stars. Comparing to figure (14) scattering becomes important in low density and high temperature regions. In that regions the electrons are mainly outside the field of ions. Hence, with decreasing particle density the true absorption

5.4 MEAN FREE PATH METHODS

becomes unimportant.

The Rosseland and Planck averaged scattering mean appearing in figure (14) have been obtained by inserting the Klein-Nishina formula (5.3.2) in (5.4.2) and (5.4.5). The Thomson cross section is not influenced by the Rosseland and Planck weighting functions. As one can recognises the scattering contribution by the Klein-Nishina formula decreases by increasing temperature starting from $k_B T \geq 2$ keV. This is due to the pair production at high temperatures [Landau87].

One can see in figure (14) that the Thomson scattering Σ_{Th} is quite good approximation for temperatures below 3 keV. Because of the diagonal Thomson scattering kernel the transport cross section is equal to the total cross section Σ_{tot} [Pomraning73]

$$\Sigma_{tr} = \Sigma_{tot} = \Sigma'_a + \Sigma_{Th}. \qquad (5.4.10)$$

The influence of the free-free, bound-free and Thomson scattering absorption on the mean free path is shown in figure (15).

5 RADIATION CROSS SECTIONS

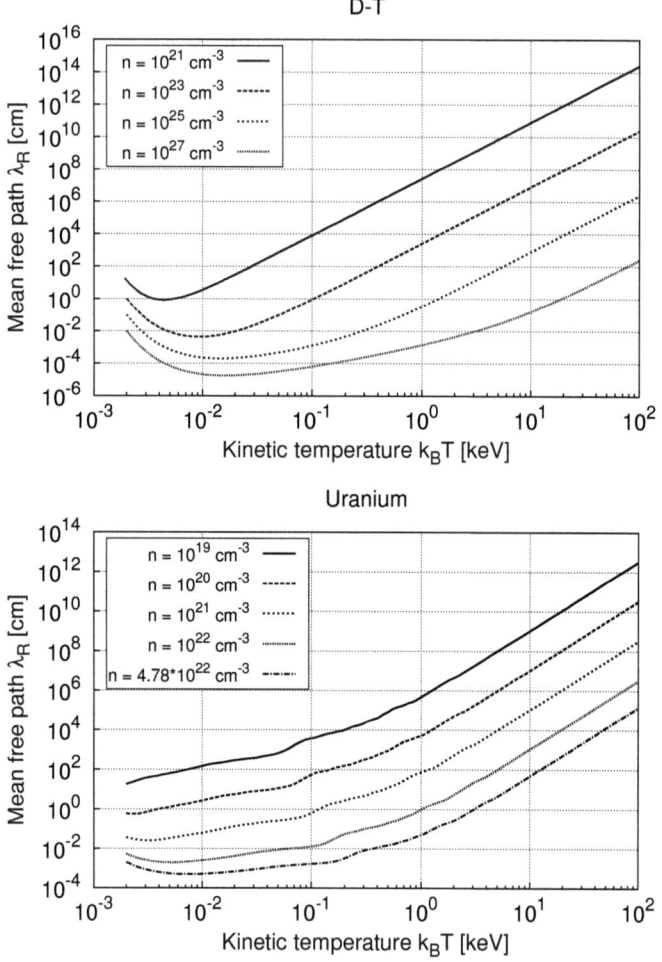

Figure 13: Rosseland mean free path of photons in a plasma consisting of D-T and uranium ions depending on temperature and density. Free-free transitions have been considered only. $n_{tot} = 4.78 \times 10^{22}$ cm^{-3} is the normal particle density of uranium.

5.4 MEAN FREE PATH METHODS

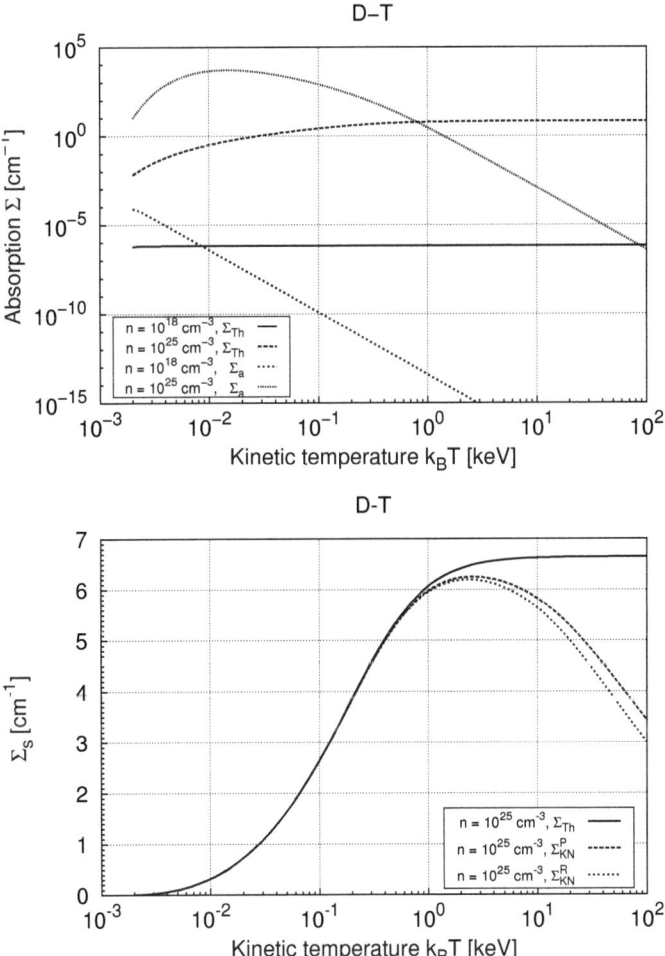

Figure 14: Scattering and absorption cross section in a D-T plasma. Scattering becomes important in optical thin regions. That means, in regions with low particle density and high temperatures. Σ_{KN}^P and Σ_{KN}^R is the Planck and Rosseland averaged Klein-Nishina formula, respectively.

5 RADIATION CROSS SECTIONS

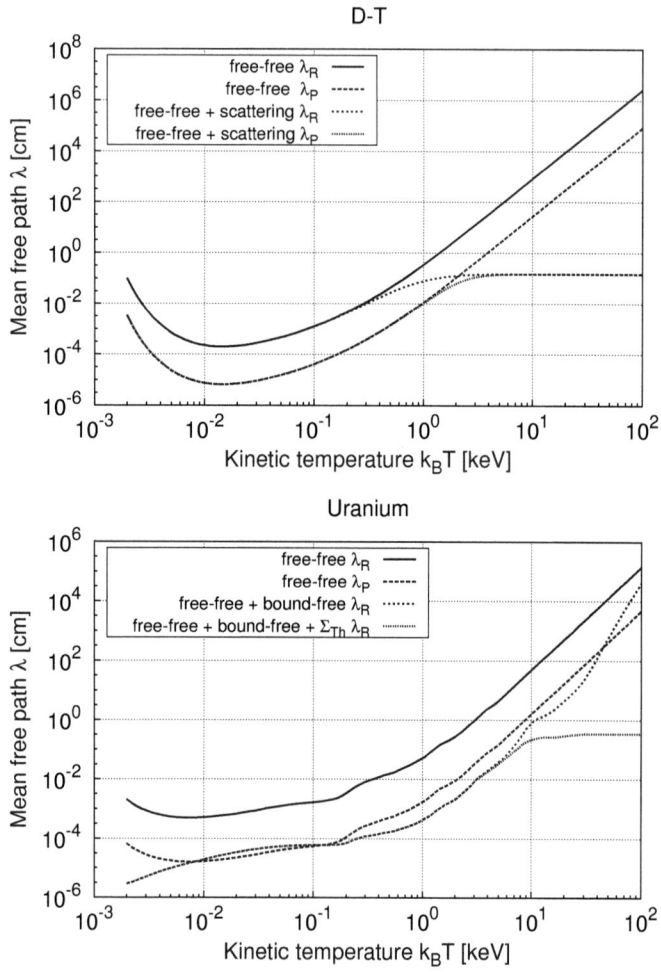

Figure 15: Rosseland averaged mean free path in a D-T plasma with particle density $n_{tot} = 10^{25}$ cm^{-3} and uranium plasma at normal particle density $n_{tot} = 4.78 \times 10^{22}$ cm^{-3} including scattering contributions from Thomson scattering.

6 Aspects of conduction

Within this section the mechanism of energy transport without carriage of particles is discussed. That physical instrument is called conduction. Depending on temperature there are two different processes exist. At temperatures below approximately 200 eV heat conduction effects dominate, at temperatures above radiation heat conduction plays a prevailing role. Radiation heat conduction significantly reduce the problem of solving the radiation transport equation approximately. That concept is derived in section (6.2). Schematically the concept of conduction is shown in figure (16).

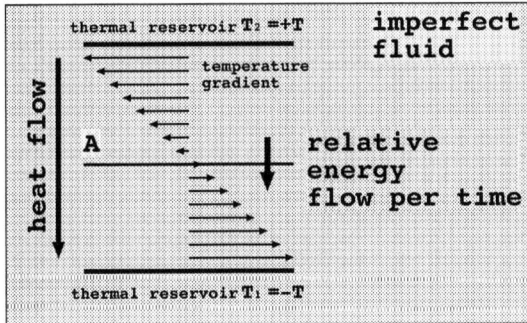

Figure 16: A model of heat flow in a slab geometry.

6.1 Heat conduction

Mostly, the heat flux is introduced by an application of Fourier's law. In that case the heat flux is separated in a coefficient κ^{th} and the temperature gradient. κ^{th} is called the heat conductivity coefficient

$$S_F^{th} = -\kappa^{th}\frac{\partial T}{\partial r}. \tag{6.1.1}$$

For the current investigations the heat conductivity is calculated by the Lorentz gas model. In that model the electrons do not interact and the ions are at rest [Spitzer62]. Due to the charge of electrons the conductivity coefficient differs from those in a gas of neutral particles. This is because of the Coulomb interaction of charged particles. Spitzer [Spitzer62, Spitzer53] propounds a heat conductivity coefficient of

6 ASPECTS OF CONDUCTION

$$\kappa^{th} = \kappa_0^{th} \frac{(k_B T)^{5/2}}{\ln \Lambda}, \qquad (6.1.2)$$

where

$$\kappa_0^{th} = \delta(T,Z) 20 \left(\frac{2}{\pi}\right)^{3/2} \frac{(4\pi\epsilon_0)^2 k_B}{m_e^{1/2} e^4 Z} \qquad (6.1.3)$$

e is the electric charge and ϵ_0 the electric field constant. The conductivity coefficient κ^{th} (6.1.2) is derived for electron-ion interactions. For low Z-medium the electron-electron interaction becomes important and reduce the conductivity [Atzeni04]. This is considered by the scaling parameter $\delta(T,Z)$. $\delta(T,Z)$ is tabulated in [Spitzer53]. A useful continuous approximation of the function $\delta(T,Z)$ for a ICF pellet corona is found in [Duderstadt82]

$$\delta(T,Z) \approx \delta(Z) = \frac{0.095(Z + 0.24)}{1 + 0.24 Z}. \qquad (6.1.4)$$

Z	δ	κ_0^{th} [10^{12} erg/μs/cm/K]
1	0.095	8.52584×10^{-6}
92	0.37967	3.70367×10^{-7}
94	0.38	3.62802×10^{-7}

Table 9: κ_0^{th} and δ tabulated for hydrogen, uranium and plutonium.

The approach (6.1.4) has been applied to the current investigations. Values for δ and κ_0^{th} are presented in table (9). The quantity $\ln \Lambda$ is called Coulomb logarithm. This term is obtained by Coulomb scattering investigations of an electron with an ion. By taking $\ln \Lambda$ into account small angle collisions are more effective than large angle collisions [Zel'dovich66, p. 419]. The following expression of $\ln \Lambda$ is taken from Spitzer [Spitzer62] and has been implemented into the program

$$\ln \Lambda = 33.825 + \ln \left(\frac{1}{ZZ_f} \frac{(k_B T)^{3/2}}{\sqrt{n_e}} \right). \qquad (6.1.5)$$

$k_B T$ is given in keV and n_e in cm^{-3}. Spitzer [Spitzer62] defines by Z_f the nuclear charge of the field particles and by Z the nuclear charge of the test particles. Z_f is identified as the nuclear charge of the ions and $|Z| = 1$ as charge of the electrons. The Coulomb logarithm for hydrogen, plutonium and uranium are presented in figure (17). Other approaches of $\ln \Lambda$ are derived in [Atzeni04, Pfalzner06, Duderstadt82].

6.1 HEAT CONDUCTION

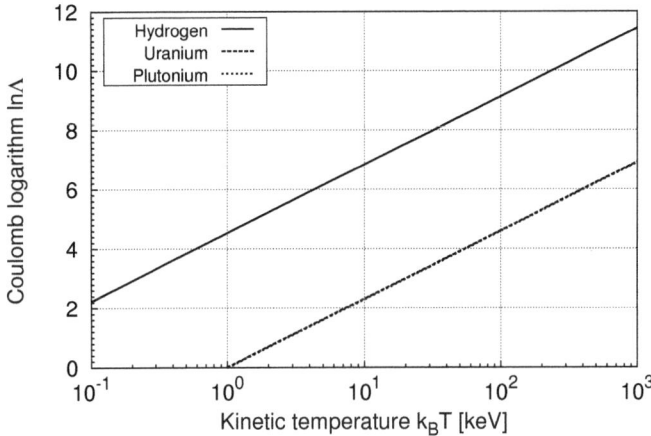

Figure 17: Coulomb logarithm for hydrogen, uranium and plutonium. The Coulomb logarithm is a slowly varying function in temperature.

In theoretical discussions κ^{th} is often assumed to have a power dependence on temperature and density. $\kappa^{th} = \kappa_0^{th} T^\nu \rho^{-\mu}$, where μ and ν are some arbitrary real constants. κ_0^{th} is a conduction coefficient at some initial conditions in temperature and density. The choice of μ and ν depends on the physical problem. For electronic heat conduction and fully ionised plasma $\nu = 5/2$ and $\mu = 0$ and for radiation heat conduction $\nu = 4 - 6$ and $\mu = 1 - 2$ [Pakula85]. That concept has been discussed by several authors. In [Pakula85] the problem of heat transfer by nonlinear conduction in dense matter is solved by a self-similar ansatz.

The ansatz (6.1.1) is valid for small temperature gradients only. To simplify matters this ansatz is used for steep gradients in temperature too. In this way the thermal flux can exceed its physical limit of a free streaming flux with some upper maximum velocity at steep gradients in temperature as one can see in figure (18). Flux limiters have been studied widely in the literature to correct this unphysical behaviour [Duderstadt82, Pfalzner06]. A natural way is to limit the flux, determined by a Fourier ansatz, to its physical maximum at increasing temperature gradient. This can be done by the functional approach [Pfalzner06]

$$S^{th} = \left(\frac{1}{S_F^{th}} + \frac{1}{S_L^{th}} \right)^{-1}. \tag{6.1.6}$$

55

6 ASPECTS OF CONDUCTION

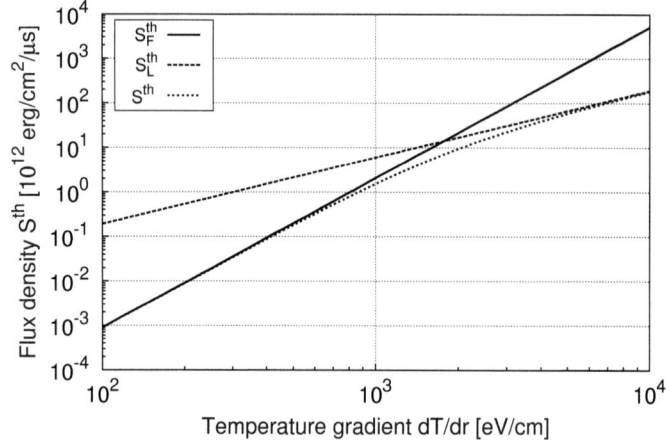

Figure 18: Behaviour of the thermal flux limiter by an increasing gradient in temperature for the case of hydrogen with an initial particle density of $n_{tot} = 10^{24}$ cm^{-3}. S_F^{th} is the flux density given by a Fourier law (equation (6.1.1)), S_L^{th} is the free streaming limit reduces by the flux limiter f_L. S^{th} is the adjusted flux.

$S_L^{th} = f_L S_s^{th}$. S_s^{th} is the free streaming limit

$$S_s^{th} = v_e n_e k_B T. \qquad (6.1.7)$$

v_e is the electron velocity obtained from a Maxwell Boltzmann distribution and n_e the electron density obtained by the equation of Saha (4.3.1). f_L is called the thermal flux limiter. The value of $f_L \approx 0.08$ has been determined to be acceptable for inertial confinement fusion [Atzeni04].

6.2 Radiation diffusion approximation

The transport of radiation is given by equation (3.2.5). As mentioned earlier solving this equation is a complex task. Approximation methods like the diffusion approximation are of widely use in studying radiative transfer problems. When the plasma is optical thick, as discussed in a previous section, radiation is absorbed and re-emitted in small spatial steps. The radiation is transported diffusively in this case. The radiation transport process is well described by a diffusion theory. The basis for this theory is the determination of the moments of the radiation transport equation (3.2.5).

6.2.1 Momentum equations

The intensity $I_\nu = I_\nu(\mathbf{r}, \mathbf{\Omega}, t)$ depends on seven independent variables. The basic idea of the diffusion approximation is the reduction of the complexity by a series expansion of the intensity in $\mathbf{\Omega}$. The expansion is truncated after the linear term. In that way the intensity reads

$$I_\nu(\mathbf{r}, \mathbf{\Omega}, t) = \frac{c}{4\pi} U_\nu(\mathbf{r}, t) + \frac{3}{4\pi} \mathbf{\Omega} \cdot \mathbf{S}_\nu(\mathbf{r}, t). \quad (6.2.1)$$

This ansatz is valid for small deviations in isotropy only. Hence, $cU_\nu \gg |\mathbf{S}_\nu|$. The derivation of the diffusion equation requires the zeroth and first moment of the radiation transport equation. The zeroth moment or energy balance equation is obtained by an angle integration of (3.2.5).

$$\frac{\partial}{\partial t} U_\nu + \nabla \cdot \mathbf{S}_\nu = \Sigma'_a \left(4\pi B_\nu(\theta) - cU_\nu \right) - c\Sigma_s U_\nu + c \int_0^\infty d\nu' \frac{\nu}{\nu'} \Sigma_{s0} \left(\nu' \to \nu \right) U_{\nu'}$$
$$+ \frac{U_\nu c^4}{8\pi h} \int_0^\infty d\nu' \left(\frac{1}{\nu'\nu^2} \Sigma_{s0}(\nu' \to \nu) - \frac{1}{\nu'^3} \Sigma_{s0}(\nu \to \nu') \right) U_{\nu'} \quad (6.2.2)$$
$$+ \frac{3}{8\pi} \frac{c^2}{h} \mathbf{S}_\nu \cdot \int_0^\infty d\nu' \left(\frac{1}{\nu'\nu^2} \Sigma_{s1}(\nu' \to \nu) - \frac{1}{\nu'^3} \Sigma_{s1}(\nu \to \nu') \right) \mathbf{S}_{\nu'}.$$

In (3.2.5) it is assumed, that the radiative source is Planck distributed. Similarly, one obtains the first moment or momentum balance equation by a multiplication with $\mathbf{\Omega}$ and an angle integration

6 ASPECTS OF CONDUCTION

$$\begin{aligned}
\frac{1}{c}\frac{\partial}{\partial t}\mathbf{S}_\nu + c\nabla \cdot \hat{\mathbf{1}}\frac{U_\nu}{3} &= -\left(\Sigma'_a + \Sigma_s\right)\mathbf{S}_\nu + \int_0^\infty d\nu' \frac{\nu}{\nu'}\Sigma_{s1}\left(\nu' \to \nu\right)\mathbf{S}_{\nu'} \\
&+ \frac{1}{8\pi}\frac{c^3}{h}U_\nu \int_0^\infty d\nu' \left(\frac{1}{\nu^2\nu'}\Sigma_{s1}(\nu' \to \nu) - \frac{1}{\nu'^3}\Sigma_{s1}(\nu \to \nu')\right)\mathbf{S}_{\nu'} \\
&+ \frac{1}{8\pi}\frac{c^2}{h}\mathbf{S}_\nu \int_0^\infty d\nu' \left(\frac{1}{\nu^2\nu'}\Sigma_{s0}(\nu' \to \nu) - \frac{1}{\nu'^3}\Sigma_{s0}(\nu \to \nu')\right)U_{\nu'}.
\end{aligned} \quad (6.2.3)$$

$\hat{\mathbf{1}}$ is the unit tensor. $\Sigma_{s0}(\nu' \to \nu)$ and $\Sigma_{s1}(\nu' \to \nu)$ is the zeroth and first moment of the differential scattering contribution $\Sigma_s(\nu' \to \nu, \mathbf{\Omega}' \cdot \mathbf{\Omega})$. Σ_s means the frequency and angle integrated scattering coefficient. In contrast to the frequency integrated quantities U (3.1.8) and $\mathbf{S}$ (3.1.9) the similar frequency dependent quantities U_ν and $\mathbf{S}_\nu$ have been used while deriving (6.2.2) and (6.2.3).

Of widely use in the literature is the case of a medium with underlying isotropic scattering. In that point the first and second moment as well as the diffusion equation are simplified significantly. The differential scattering kernel reads [Pomraning73]

$$\Sigma_s(\nu' \to \nu, \mathbf{\Omega} \cdot \mathbf{\Omega}') = \frac{1}{4\pi}\Sigma_s(\nu')\delta(\nu - \nu'). \quad (6.2.4)$$

In that way the remaining part of the Legendre expansion is

$$\Sigma_{s0}(\nu' \to \nu) = \Sigma_s(\nu')\delta(\nu - \nu'). \quad (6.2.5)$$

All higher moments are zero. The equations (6.2.2) and (6.2.3) simplify significantly to

$$\frac{\partial}{\partial t}U_\nu + \nabla \cdot \mathbf{S}_\nu = \Sigma'_a\left(4\pi B_\nu(\theta) - cU_\nu\right) \quad (6.2.6)$$

$$\frac{1}{c}\frac{\partial}{\partial t}\mathbf{S}_\nu + c\nabla \cdot \hat{\mathbf{1}}\frac{U_\nu}{3} = -\left(\Sigma'_a + \Sigma_s\right)\mathbf{S}_\nu. \quad (6.2.7)$$

The law of Fick postulates, that the radiation flux is related to the radiation energy by[6]

$$\mathbf{S}_\nu = cD_\nu \nabla U_\nu. \quad (6.2.8)$$

D_ν is called the diffusion coefficient and defined by

---

[6]This follows immediately by assuming a weak time dependency in (6.2.7) and accepting $D_\nu = (3\left(\Sigma'_a + \Sigma_s\right))^{-1}$.

6.2 RADIATION DIFFUSION APPROXIMATION

$$D_\nu = \frac{1}{3\left(\Sigma'_a + \Sigma_s\right)}. \qquad (6.2.9)$$

Inserting (6.2.8) in (6.2.6) results in the simplified diffusion equation

$$\frac{1}{c}\frac{\partial}{\partial t}U_\nu - \nabla \cdot (D_\nu \nabla U_\nu) = \Sigma'_a \left(\frac{4\pi}{c}B_\nu(\theta) - U_\nu\right). \qquad (6.2.10)$$

6.2.2 Radiation heat conduction - equilibrium diffusion approximation

Assuming a weak time and spatial dependency in U_ν. Than the derivative in time and spatial coordinates in equation (6.2.10) can be neglected. This immediately gives the radiation energy density in the limit of the *equilibrium diffusion approximation*

$$U_\nu = \frac{4\pi}{c}B_\nu(\theta). \qquad (6.2.11)$$

The flux reads

$$\mathbf{S}_\nu = \frac{c}{3\left(\Sigma'_a + \Sigma_s\right)}\nabla B_\nu. \qquad (6.2.12)$$

Taking only frequency integrated quantities into account one can apply the results from section (5). In that way U and S in the one-dimensional case read

$$U = a_{rad}T^4 \qquad (6.2.13)$$
$$S = 4a_{rad}c\frac{\lambda_R}{3}\frac{dT}{dr}. \qquad (6.2.14)$$

In analogy to the definition of the heat conduction coefficient (6.1.2) the radiation heat conduction coefficient

$$\kappa^{rad} = 4a_{rad}c\frac{\lambda_R}{3} \qquad (6.2.15)$$

is introduced. The relevance of pure heat conduction and radiation heat conduction depending on temperature and density is shown in figure (19). One recognises, that the pure heat conduction is relevant at small temperatures only. One can show, that contributions of pure heat conduction become neglectable at lower particle densities. For later purposes the heat diffusivity χ

6 Aspects of Conduction

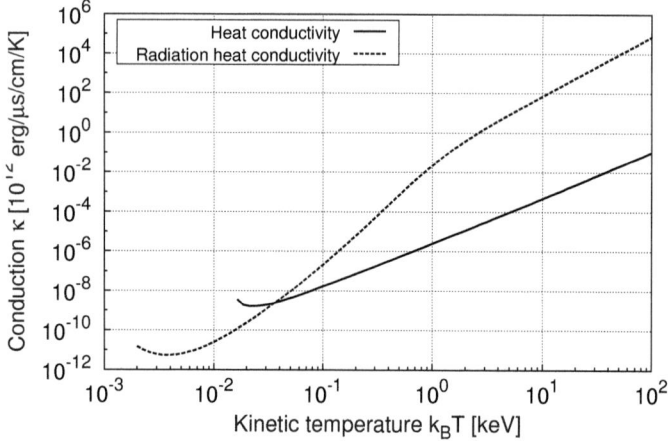

Figure 19: Heat conduction and radiation heat conduction coefficient depending on temperature for a D-T plasma having an initial particle density $n_{tot} = 10^{25}$ cm^{-3}.

$$\chi = \frac{\kappa}{\rho c_V}, \qquad (6.2.16)$$

is introduced, where κ is defined by (6.1.2) or (6.2.15). The specific heat is given by (3.4.16) or (3.4.27). χ is responsible for internal time step calculations in **STEALTH**. In the literature, χ is of widely use in dimensionless analysis problems [Landau87]. The behaviour of the thermal diffusivity depending on temperature and density is shown in figure (20). The radiation heat diffusivity is almost always several orders higher than the pure heat diffusivity. This follows from the conductivity contributions. If the specific heat is constant the heat diffusivity is proportional to the heat conductivity. The specific heat depending on temperature (see equation 3.4.27) increases with increasing temperature. Hence, the influence of the conductivity coefficient attenuates with increasing temperature and the diffusivity shapes a saturation plateau. Refer to figure (19).

6.2.3 Flux limiters

As in the case of pure heat conduction the flux ansatz using a Fourier law is valid for small temperature gradients only. This condition is fulfilled in optical thick systems. Due to the

6.2 RADIATION DIFFUSION APPROXIMATION

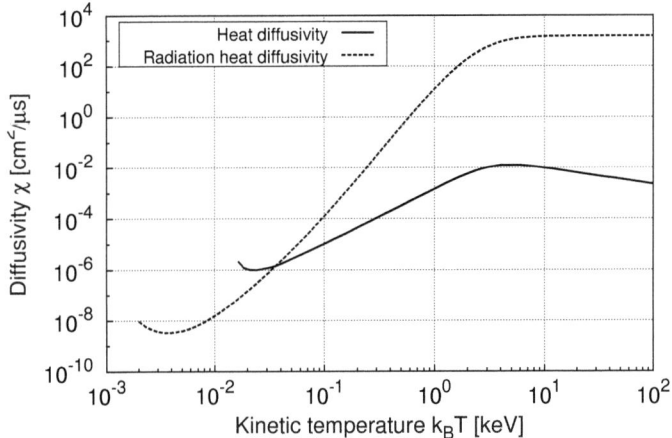

Figure 20: Thermal diffusivity depending on temperature calculated using a heat conduction (6.1.2) and a radiation heat conduction approach (6.2.15) for a D-T plasma having an initial particle density $n_{tot} = 10^{25}$ cm^{-3}.

numerical difficulties and computational costs in solving the equation of radiative transfer (3.2.5) that ansatz has been applied to regions with steep gradients in temperature too - optical thin regime. Following the same argumentation in section (6.1) one has to guarantee, that the radiative flux is limited to its physical maximum cU at steep temperature gradients, otherwise the flux violates the causality principle. This leads to the concept of radiative flux limiters, which is studied widely in the literature.

For the current investigations a flux limiter D_n is introduced by

$$D_n = \left(\left(\frac{3}{\lambda_R}\right)^n + \left(\frac{1}{U}\frac{\partial}{\partial r}U\right)^n\right)^{-1/n}. \qquad (6.2.17)$$

Lund and Wilson [Wilson80] proposed the flux limiter D_1. It is shown, that in the optical thick limit, where the diffusion approximation holds, this limiter is accurate in zeroth order only compared to transport theory [Morel00]. Taking $n = 2$ a first order accuracy in the optical thick limit is reached. That flux limiter, D_2, has been proposed by Larsen [Olson00]. In limit of $\lim n \to \infty$ the flux limiter reads

6 ASPECTS OF CONDUCTION

$$D_{max} = \left(\max\left[\frac{3}{\lambda_R}, \frac{1}{U}\frac{\partial}{\partial r}U\right]\right)^{-1}, \qquad (6.2.18)$$

but no investigations regarding the accuracy of this limiter has been given yet [Olson00]. We restrict ourselves to the case D_1. The influence of higher order D_n to our results needs to be studied in future investigations. Olson et al. [Olson00] derived a new flux limiter, called P$_{1/3}$ based on a simple combination of the diffusion and P_1 equation. Simmon et al. [Simmons00] investigated this modified P$_1$ equation in a linearised analysis. It is shown that $P_{1/3}$ gives the correct propagation speeds in both the optical thick and thin limit. Morel [Morel00] derived an asymptotic solution to the transport equation to first order. Comparing this result it is shown, that Olson's $P_{1/3}$ and the diffusion equation are correct to first order too, but Wilson's flux limiter D_1 is correct to zeroth order only. Other forms of flux limiters have been studied by Pomraning et al. [Sanchez91, Pomraning81] based on a generic form of Fick's law. Beside the *ad hoc* introduction of flux limiters Levermore [Levermore79] developed an intrinsically flux-limited diffusion theory based on a Chapman-Enskog procedure.

7 Multigroup approach

In section (6) the concepts of heat conduction and radiation heat conduction were introduced assuming material and radiation in local thermodynamical equilibrium. Often, the equilibrium condition is not fulfilled or it is of particular interest to study time-dependent, non-equilibrium and frequency dependent radiative transfer. As mentioned earlier solving the full equation of radiative transfer is a complex and cost intensive task. To deal with frequency-dependent variables in the radiative transport equation one introduces the multigroup method. Within this method the frequency spectrum is splitted in frequency groups. All photons belonging to one frequency group have the same frequency-independent averaged properties, which are governed for that group. This method is at least accurate for coarsened frequency groups. The results will become better by using more fine structured frequency groups, but this will become computationally costly too.

Pritzker et al. [Pritzker76] published six-energy group constants of the absorption coefficient (bound-bound, bound-free and free-free absorption included), the scattering coefficient as well as moments of the inscattering term. Considering an uranium plasma they assumed having a system in local thermal equilibrium between temperatures of 0.1 keV and 10 keV at solid state density. Their numerical work has been repeated for some checks. Beside their computational methods analytical solutions in the low photon energy limit are presented within this section. Low photon energies means energies much smaller than the electron mass at rest. This limit is introduced by the expression $\gamma = h\nu/m_e c^2 \ll 1$, where $h\nu$ is the photon energy, m_e the electron mass and c the speed of light. In a wide energy range the presented results are in precise agreement with those gained by Pritzker et al. [Pritzker76].

Starting from the general description neglecting induced scattering terms a multigroup presentation of the equation of radiative transfer valid for local thermal equilibrium is derived. Afterwards, multigroup approximations for absorption and scattering terms are given. Concrete results are obtained using the photon energy group splitting as given by [Pritzker76].

7 MULTIGROUP APPROACH

7.1 Group integrated quantities

The radiation transport equation neglecting induced scattering terms reads [Pomraning73]

$$\frac{1}{c}\frac{\partial I_\nu(\mathbf{\Omega})}{\partial t} + \mathbf{\Omega} \cdot \nabla I_\nu(\mathbf{\Omega}) = \Sigma'_a(\nu)\left(B_\nu(\theta) - I_\nu(\mathbf{\Omega})\right) - \Sigma_s(\nu)I_\nu(\mathbf{\Omega}) \\ + \int_0^\infty d\nu' \int_{4\pi} d\mathbf{\Omega}' \frac{\nu}{\nu'}\Sigma_s(\nu' \to \nu, \mathbf{\Omega}' \cdot \mathbf{\Omega})I_{\nu'}(\mathbf{\Omega}'). \quad (7.1.1)$$

The spatial and time dependencies in I are suppressed. Following Pritzker et al. [Pritzker76] the photon spectrum is divided into six energy groups. The energy groups are presented in table (10). The group identifier is g.

g	1	2	3	4	5	6
$h\nu_g$	5.00e+02	6.40e+01	1.60e+01	4.00e+00	1.00e+00	2.5e-01
$h\nu_{g+1}$	6.40e+01	1.60e+01	4.00e+00	1.00e+00	2.50e-01	1.0e-04

Table 10: Photon energy groups. The photon energy is given in keV. $h\nu_g$ means the upper photon energy boundary and $h\nu_{g+1}$ means the lower photon energy boundary.

The integration of (7.1.1) over the energy groups leads to

$$\left(\frac{1}{c}\frac{\partial}{\partial t} + \mathbf{\Omega} \cdot \nabla\right) I_g(\mathbf{\Omega}) = -\left(\Sigma'_{ag} + \Sigma_{stg}\right) I_g(\mathbf{\Omega}) + Q_g + \int_g d\nu\, \mathcal{C} \quad (7.1.2)$$

where $\int_g d\nu\,[\dots]$ is a shortened term for $\int_{\nu_{g+1}}^{\nu_g} d\nu\,[\dots]$. $\mathcal{C}$ indicates the inscattering contribution

$$\mathcal{C} = \int_0^\infty d\nu' \int_{4\pi} d\mathbf{\Omega}' \frac{\nu}{\nu'}\Sigma_s(\nu' \to \nu, \mathbf{\Omega}' \cdot \mathbf{\Omega})I_{\nu'}(\mathbf{\Omega}') \quad (7.1.3)$$

and will be considered later. The variables $\Sigma'_{ag}, I_g, B_g, Q_g, \Sigma_{stg}$ denote the frequency group (see table (10)) integrated quantities of the macroscopic absorption coefficient, specific intensity, black body radiation, emission source and the macroscopic scattering coefficient, respectively. Formally, those quantities are given by [Pomraning73]

$$I_g(\Omega) = \int_g d\nu\, I_\nu(\Omega) \tag{7.1.4}$$

$$\Sigma'_{ag} = \left(\int_g d\nu\, \Sigma'_a (B_\nu(\theta) - I_\nu)\right) \times \left(\int_g d\nu\, (B_\nu(\theta) - I_\nu)\right)^{-1} \tag{7.1.5}$$

$$\Sigma_{stg} = \frac{1}{I_g}\int_g d\nu\, \Sigma_s(\nu) I_\nu \tag{7.1.6}$$

$$Q_g = \int_g d\nu\, \Sigma'_a(\nu) B_\nu(\theta) \tag{7.1.7}$$

$$B_g(\theta) = \int_g d\nu\, B_\nu(\theta), \tag{7.1.8}$$

where $\theta = k_B T$ is the kinetic temperature. In a local thermal equilibrium regime Σ'_{ag} and Σ_{stg} can be approximated by their Planck averaged quantities. In this case Σ'_{ag} and Σ_{stg} read

$$\Sigma'_{ag} = \frac{1}{B_g}\int_g d\nu\, \Sigma'_a(\nu) B_\nu(\theta) \tag{7.1.9}$$

$$\Sigma_{stg} = \frac{1}{B_g}\int_g d\nu\, \Sigma_s(\nu) B_\nu(\theta) \tag{7.1.10}$$

Generally, the radiation source is given by the ratio of the emission and absorption coefficient. In local thermal equilibrium the source is of Planckian type. Therefore one has for the group emission source

$$Q_g = \int_g d\nu\, \Sigma'_a(\nu) B_\nu(\theta) = \Sigma'_{ag} B_g. \tag{7.1.11}$$

7.2 Group integrated Planck function

Inserting the Planck function in (7.1.8) leads to

$$B_g = \frac{2\theta^4}{c^2 h^3}\int_{u_{g+1}}^{u_g} du\, u^3 (\exp(u) - 1)^{-1} \tag{7.2.1}$$

where u_g and u are defined as $u_g = h\nu_g/\theta$ and $u = h\nu/\theta$. In contrast to the numerical discussion of (7.1.8) in [Pritzker76] an analytical solution of this integral is possible. The integral is of type $\mathcal{I}_3$ as given in the appendix (A.3.13). Therefore the solution is given by

$$B_g = \frac{2\theta^4}{c^2 h^3}\mathcal{I}_3(u_{g+1}, u_g). \tag{7.2.2}$$

7 MULTIGROUP APPROACH

The results for the six group Planck spectrum B_g in units of keV/cm²s are presented in table (11).

g/θ	1.00e-01	3.16e-01	1.00e+00	3.16e+00	1.00e+01
1	9.329e-243	2.911e-52	1.387e+09	4.852e+28	2.246e+35
2	4.277e-36	4.428e+12	1.758e+28	4.834e+33	1.593e+36
3	9.228e+14	2.585e+27	8.230e+31	1.426e+34	2.194e+35
4	1.951e+26	1.161e+30	1.149e+32	1.251e+33	5.657e+33
5	1.443e+28	8.352e+29	6.923e+30	2.887e+31	9.936e+31
6	5.801e+27	3.802e+28	1.490e+29	5.025e+29	1.623e+30

Table 11: Six-group Planck spectrum B_g in units of keV/cm²s. g stands for the group index and θ is the temperature keV. No contributions to B_g are given for high energy groups and small temperatures. In that case Pritzker et al. [Pritzker76] use a transformed presentation for $B_\nu(\theta)$ called $B_\nu^*(\theta)$. To avoid numerical problems while evaluating group constants B_g^* is defined by $B_\nu^*(\theta) = \exp(u_{g+1})B_\nu(\theta)$. In the present analytical considerations no such transformations are necessary.

7.3 Absorption cross section

The absorption cross section corrected for induced scattering effects reads [Zel'dovich66]

$$\Sigma'_a = \Sigma_a \left(1 - \exp(-u)\right). \tag{7.3.1}$$

Σ'_a is discussed in [Fiedler07a] and includes the contributions of bound-free and free-free transitions corrected for induced effects [Zel'dovich66]. Generally, Σ'_a includes contributions from bound-bound states too, but these have a small effect at high temperatures. For calculation of Σ'_a the new program system enrico [Fiedler07b] was developed. The result is given in table (12). In a pure free-free absorbing system an analytical solution for the multigroup absorption coefficient is possible. The free-free absorption coefficient is given by [Fiedler07a]

$$\Sigma_{ff}(\nu) = 0.767 \times 10^{-47} \frac{\bar{r}^3 n_i^2}{u^3 \theta^{7/2}} \quad [\text{cm}^{-1}], \tag{7.3.2}$$

where n_i is the ion density and $\bar{r} = n_e/n_i$ the averaged ionisation number. Using (7.1.11) one finds the Planck averaged free-free group absorption cross section

$$\Sigma'_{ag} = 0.767 \times 10^{-47} \frac{1}{B_g} \frac{2}{h^3 c^2} \bar{r}^3 n_i^2 \sqrt{\theta} \left(\exp(-u_{g+1}) - \exp(-u_g)\right) \quad [\text{cm}^{-1}]. \tag{7.3.3}$$

The results are given in table (13). Advanced concepts in studying multigroup absorption cross sections have been reported by [Li09].

g/θ	1.00e-01	3.16e-01	1.00e+00	3.16e+00	1.00e+01
1	4.699e-03	3.101e-02	1.789e-01	3.634e+01	1.604e+00
2	3.933e-02	2.220e-01	2.321e+02	1.577e+02	2.719e+00
3	1.699e+01	7.831e+02	5.734e+03	1.397e+02	6.099e+00
4	2.564e+03	1.809e+04	2.084e+03	3.926e+02	8.552e+01
5	5.346e+04	1.488e+04	1.361e+04	5.819e+03	1.369e+03
6	1.531e+05	1.735e+05	2.793e+05	1.232e+05	2.886e+04

Table 12: Six-group Planck averaged absorption cross section Σ'_{ag} for uranium at solid state density obtained from (7.1.11) by a numerical calculation using the program system `enrico` [Fiedler07b]. The absorption contribution contains bound-free and free-free transitions. The results are different from those obtained by Pritzker et al [Pritzker76]. They use slightly different ionization energies and include first order corrections to absorption contributions from excited atomic states [Pritzker75].

g/θ	1.00e-01	3.16e-01	1.00e+00	3.16e+00	1.00e+01
1	4.438e-04	2.671e-03	1.403e-02	1.783e-02	9.404e-03
2	2.801e-02	1.634e-01	7.765e-01	7.072e-01	1.598e-01
3	1.693e+00	8.691e+00	2.700e+01	1.045e+01	2.714e+00
4	8.558e+01	2.569e+02	3.691e+02	1.929e+02	5.270e+01
5	2.091e+03	3.477e+03	7.200e+03	3.655e+03	9.016e+02
6	5.813e+04	1.015e+05	1.800e+05	8.178e+04	1.933e+04

Table 13: Six-group Planck mean pure free-free absorption cross section for uranium at solid state density (7.3.3).

7.4 Scattering cross section

The scattering contribution as appearing in (7.1.2) is obtained by using the angle integrated Compton cross section [Pomraning73]

$$\Sigma_s(\nu) = \frac{3}{4}\Sigma_{Th}\left\{\left(\frac{1+\gamma}{\gamma^3}\right)\left[\frac{2\gamma(1+\gamma)}{1+2\gamma} - \ln(1+2\gamma)\right] + \right. \\ \left. + \frac{1}{2\gamma}\ln(1+2\gamma) - \frac{1+3\gamma}{(1+2\gamma)^2}\right\} \quad (7.4.1)$$

7 MULTIGROUP APPROACH

where $\Sigma_{Th} = 0.665 \times 10^{-24} \, n_e$ is the macroscopic Thomson cross section and $\gamma = h\nu/m_e c^2$. Using (7.1.10) and the abbreviation $\gamma_g = h\nu_g/m_e c^2$ one obtains the Planck averaged scattering contribution

$$\Sigma_{stg} = \frac{\Sigma_{Th}}{B_g} \left(\frac{3(m_e c^2)^4}{2c^2 h^3}\right) \int_{\gamma_{g+1}}^{\gamma_g} d\gamma \frac{\gamma^3}{(\exp(\gamma m_e c^2/\theta) - 1)} \times \\ \times \left\{\left(\frac{1+\gamma}{\gamma^3}\right)\left[\frac{2\gamma(1+\gamma)}{1+2\gamma} - \ln(1+2\gamma)\right] + \frac{1}{2\gamma}\ln(1+2\gamma) - \frac{1+3\gamma}{(1+2\gamma)^2}\right\} \quad (7.4.2)$$

The integration is carried out numerically. The results are given in table (14).

g/θ	1.00e-01	3.16e-01	1.00e+00	3.16e+00	1.00e+01
1	8.140e-01	8.136e-01	8.121e-01	8.067e-01	7.851e-01
2	9.420e-01	9.413e-01	9.385e-01	9.260e-01	8.831e-01
3	9.843e-01	9.832e-01	9.781e-01	9.635e-01	9.574e-01
4	9.956e-01	9.937e-01	9.899e-01	9.887e-01	9.884e-01
5	9.982e-01	9.973e-01	9.971e-01	9.971e-01	9.970e-01
6	9.996e-01	9.994e-01	9.994e-01	9.994e-01	9.994e-01

Table 14: Six-group Planck mean Compton cross section in Thomson units evaluated at different temperatures θ in keV. Misleadingly to the notes of Pritzker [Pritzker76] the ratios Σ_{stg}/Σ_{Th} obtained from (7.4.2) are independent from material and density. The results above are evaluated numerically while using the Gauss-Kronrod integration procedure.

In case of small photon energies, e.g. $\gamma \ll 1$, an analytic solution is derived. The scattering cross section (7.4.1) corrected to second order is [Pomraning73]

$$\Sigma_s = \Sigma_{Th}\left(1 - 2\gamma + \frac{26}{5}\gamma^2\right). \quad (7.4.3)$$

Inserting the above expression in (7.1.10) yields

$$\Sigma_{stg} = \frac{\Sigma_{Th}}{B_g}\left(\frac{2(m_e c^2)^4}{c^2 h^3}\right)\left\{\int_{\gamma_{g+1}}^{\gamma_g} d\gamma \frac{\gamma^3}{(\exp(\gamma m_e c^2/\theta) - 1)} \right. \\ \left. -2\int_{\gamma_{g+1}}^{\gamma_g} d\gamma \frac{\gamma^4}{(\exp(\gamma m_e c^2/\theta) - 1)} + \frac{26}{5}\int_{\gamma_{g+1}}^{\gamma_g} d\gamma \frac{\gamma^5}{(\exp(\gamma m_e c^2/\theta) - 1)}\right\} \quad (7.4.4)$$

Comparing the Appendix (A) Σ_{stg} is related to

$$\Sigma_{stg} = \frac{\Sigma_{Th}}{B_g} \left(\frac{2\theta^4}{c^2h^3}\right) \left\{ \mathcal{I}_3(u_{g+1}, u_g) - 2\alpha \mathcal{I}_4(u_{g+1}, u_g) + \frac{26}{5}\alpha^2 \mathcal{I}_5(u_{g+1}, u_g) \right\} \quad (7.4.5)$$

where α and u_g are given by $\alpha = \theta/m_e c^2$ and $u_g = \gamma_g/\alpha$. The results of expression (7.4.5) are presented in table (15).

g/θ	1.00e-01	3.16e-01	1.00e+00	3.16e+00	1.00e+01
1	8.309e-01	8.306e-01	8.297e-01	8.266e-01	8.191e-01
2	9.421e-01	9.414e-01	9.386e-01	9.265e-01	8.871e-01
3	9.843e-01	9.832e-01	9.780e-01	9.635e-01	9.574e-01
4	9.956e-01	9.937e-01	9.899e-01	9.887e-01	9.884e-01
5	9.982e-01	9.973e-01	9.971e-01	9.971e-01	9.970e-01
6	9.996e-01	9.994e-01	9.994e-01	9.994e-01	9.994e-01

Table 15: The table shows the six-group Planck mean Compton cross section in Thomson units evaluated at different temperatures θ in keV. The constants Σ_{stg}/Σ_{Th} obtained from (7.4.5) are independent from material and density. The results are in good agreement with those in table (14) except for the case, where $\gamma \ll 1$ is not valid.

g/θ	1.00e-01	3.16e-01	1.00e+00	3.16e+00	1.00e+01
1	2.077	2.095	2.172	2.474	4.322
2	0.010	0.011	0.016	0.053	0.456
3	0.000	0.000	0.001	0.000	0.001
4	0.000	0.000	0.000	0.000	0.000
5	0.000	0.000	0.000	0.000	0.000
6	0.000	0.000	0.000	0.000	0.000

Table 16: Accuracy of the multigroup Compton cross section in small photon energy limit. The table shows the deviation in percent from the general valid multigroup term (7.4.2). Precise agreement between the general case (table (15)) and small energy limit (table (14)) are obtained for the energy groups two to five up to temperatures of around 4 keV.

7.5 Scattering transfer cross section

The scattering part $\mathcal{C}$ of (7.1.1) reads

$$\mathcal{C}[I_\nu] = \int_{4\pi} d\Omega' \int_0^\infty d\nu' \frac{\nu}{\nu'} \Sigma_s(\nu' \to \nu, \mathbf{\Omega} \cdot \mathbf{\Omega}') I_{\nu'}(\mathbf{\Omega}') \quad (7.5.1)$$

7 MULTIGROUP APPROACH

Integration of $\mathcal{C}[I_\nu]$ over ν by splitting the integrals in different photon energy intervals gives the frequency averaged inscattering contribution

$$\mathcal{C}_g[I_\nu] = \int_{4\pi} d\Omega' \sum_{g'=gmin(g)}^{gmax(g)} \frac{1}{I_{g'}} \int_{g'} d\nu' \int_g d\nu \, \frac{\nu}{\nu'} \Sigma_s(\nu' \to \nu, \Omega' \cdot \Omega) I_{\nu'}(\Omega') \quad (7.5.2)$$

$$= \int_{4\pi} d\Omega' \sum_{g'=gmin(g)}^{gmax(g)} S_{g'g} \quad (7.5.3)$$

`gmin(g)` indicates the smallest and `gmax(g)` the highest energy group g' from which a photon is able to scatter into group g. By the summation over g' all possible photon scattering contributions from the energy group g' into the energy group g are considered. $S_{g'g}$ is the group integrated differential scattering coefficient. In local thermal equilibrium the weighting function $I_{\nu'}$ is given by Planck's function approximately. One therefore obtains for $S_{g'g}$

$$S_{g'g} = \frac{1}{B_{g'}} \int_{g'} d\nu' \frac{B_{\nu'}}{\nu'} \int_g d\nu \, \nu \Sigma_s(\nu' \to \nu, \Omega' \cdot \Omega). \quad (7.5.4)$$

The Legendre expansion of $\Sigma_s(\nu' \to \nu, \Omega' \cdot \Omega)$ reads

$$\Sigma_s(\nu' \to \nu, \Omega' \cdot \Omega) = \sum_{l=0}^{\infty} \frac{2l+1}{4\pi} \Sigma_{sl}(\nu' \to \nu) P_l(\Omega' \cdot \Omega). \quad (7.5.5)$$

where the expansion coefficients $\Sigma_{sl}(\nu' \to \nu)$ are defined by

$$\Sigma_{sl}(\nu' \to \nu) = 2\pi \int_{-1}^{1} d\mu_0 \, \Sigma_s(\nu' \to \nu, \mu_0) P_l(\mu_0) \quad (7.5.6)$$

where $\mu_0 = \Omega' \cdot \Omega$ is the scattering angle. By Using (7.5.5) and (7.5.6) in (7.5.4) one defines the Legendre ordered group integrated differential scattering coefficient

$$S_{lg'g} = \frac{2\pi}{B_{g'}} \int_{1}^{-1} d\mu_0 P_l(\mu_0) \int_{g'} d\nu' \frac{B_{\nu'}}{\nu'} \int_g d\nu \nu \Sigma_s(\nu' \to \nu, \mu_0). \quad (7.5.7)$$

l marks the Legendre order. By using the above transformations the differential scattering contribution of the radiative transfer equation reads

$$\mathcal{C}_g[I_\nu] = \sum_{l=0}^{\infty} \frac{2l+1}{4\pi} \int_{4\pi} d\Omega' P_l(\Omega' \cdot \Omega) \sum_{g'=gmin(g)}^{gmax(g)} S_{lg'g} I_{g'}(\Omega') \quad (7.5.8)$$

In the next section the contributions $S_{lg'g}$ are evaluated. Beside the numerical calculations

7.6 Moments of the group transfer scattering cross section

The differential scattering moments are given by (7.5.7). For further purposes we transform (7.5.7) by help of (3.2.2) and $\gamma = h\nu/m_e c^2$ in

$$S_{lg'g} = \frac{4\pi(m_e c^2)^3}{h^2 c^2} \frac{1}{B_{g'}} \int_{-1}^{1} d\mu_0 P_l(\mu_0) \int_{g'} d\gamma' \frac{\gamma'^2}{(\exp(\gamma' m_e c^2/\theta) - 1)} \times \\ \times \int_g d\nu \nu \Sigma_s(\nu' \to \nu, \mu_0). \qquad (7.6.1)$$

7.6.1 Group transfer scattering cross section in the Thomson limit

The Thomson differential cross section is given by [Pomraning73]

$$\Sigma_s(\nu' \to \nu, \mu_0) = \frac{3}{16\pi} \Sigma_{Th}(1 + \mu_0^2)\delta(\nu' - \nu). \qquad (7.6.2)$$

Using the above expression in (7.6.1) yields

$$\begin{aligned} S_{lg'g} &= \frac{2\pi}{B_{g'}} \int_{-1}^{1} d\mu_0 P_l(\mu_0) \int_{g'} d\nu' \frac{B_{\nu'}}{\nu'} \int_g d\nu \nu \frac{3}{16\pi} \Sigma_{Th}(1 + \mu_0^2)\delta(\nu' - \nu) \\ &= \frac{3}{8} \Sigma_{Th} \int_{-1}^{1} d\mu_0 P_l(\mu_0)(1 + \mu_0^2) \end{aligned} \qquad (7.6.3)$$

The Thomson cross section does not depend on frequency of the incident photons. No frequency shifts occur during the scattering. Hence, the outscattered photons belong to the same energy group as the inscattered photons. Owing to the definition of the Legendre polynoms one calculates

$$1 + \mu_0^2 = \frac{4}{3} P_0(\mu_0) + \frac{2}{3} P_2(\mu_0) \qquad (7.6.4)$$

and therefore

$$\begin{aligned} S_{0gg} &= \frac{\Sigma_{Th}}{2} \int_{-1}^{1} d\mu_0 P_0(\mu_0) P_0(\mu_0) = \Sigma_{Th} \\ S_{1gg} &= 0 \\ S_{2gg} &= \frac{\Sigma_{Th}}{4} \int_{-1}^{1} d\mu_0 P_2(\mu_0) P_2(\mu_0) = \frac{1}{10} \Sigma_{Th} \end{aligned} \qquad (7.6.5)$$

7 MULTIGROUP APPROACH

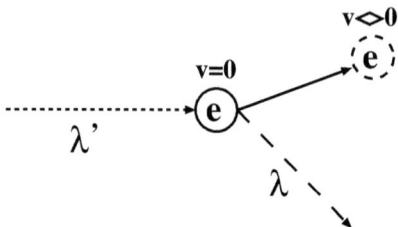

Figure 21: The Compton effect. λ' is the wave length of the incident photon. λ is the wave length of the scattered photon. v means the velocity of the electron.

or in terms of the Thomson unit $S_{0gg} = 1$, $S_{1gg} = 0$, $S_{2gg} = 1/10$. The orthogonality relation

$$\int_{-1}^{1} d\mu_0 \, P_n(\mu_0) P_m(\mu_0) = \frac{2}{2n+1} \delta_{nm}$$

has been used while evaluating the transfer cross section. All higher moments are zero.

7.6.2 Frequency shift formula

The shift of frequency of the photon scattering on an electron at rest is given by

$$\Delta\lambda = \lambda - \lambda' = \frac{h}{m_e c} (1 - \cos\phi). \tag{7.6.6}$$

In that case λ and λ' denote the wavelength of the scattered and inscattered photon, respectively ($\lambda \geq \lambda'$, $\Delta\lambda \geq 0$). Switching to the energy group notation, using the relation $\lambda = c/\nu$ and considering extremal cases only, one obtains

$$\frac{1}{\nu_g} - \frac{1}{\nu'_{g,max}} = 2\frac{h}{m_e c^2} \qquad \phi = \pi \tag{7.6.7}$$

$$\frac{1}{\nu_g} - \frac{1}{\nu'_{g,min}} = 0 \qquad \phi = 0. \tag{7.6.8}$$

Therefore one gains

7.6 Moments of the group transfer scattering cross section

$$\nu'_{max,g} = \frac{\nu_g}{1 - 2\gamma_g} \qquad (7.6.9)$$

$$\nu'_{min,g} = \nu_g, \qquad (7.6.10)$$

where γ_g is defined by $\gamma_g = h\nu_g/m_e c^2$. There is no upscattering for electrons at rest, hence $S_{lg'g} = 0$, if $g' > g$. Table (17) shows the highest frequencies $\nu'_{max,g}$ of photons which are scattered into group g. Comparing the values for $h\nu'_{max,g}$ and $h\nu_{g-1}$ one recognises $h\nu'_{max,g} < h\nu_{g-1}$[7]. Therefore one has inscattering contributions from group $g - 1$ only. $S_{lg'g} = 0$, if $g' < g - 1$. Because of no upscattering we only have to calculate S_{lg-1g} and S_{lgg}. All other contributions are zero.

Pritzker et al. [Pritzker76] take ν_{g+1} for $\nu'_{min,g}$. This is inappropriate. Their usage leads to a double counting procedure for inscattering contributions of group $g - 1$ into g and g into g. The choice of ν_g for $\nu'_{min,g}$ guarantees that only photons with frequencies greater or equal than ν_g will be considered as inscattering terms for group g. Inscattering contributions from group g into group g will be considered separately.

g	$h\nu_{g-1}$ [keV]	$h\nu_g$ [keV]	γ_g	$h\nu'_{max,g}$ [keV]	$h\nu'_{min,g}$ [keV]
1	-	5.0000e+02	-	-	-
2	5.0000e+02	6.4000e+01	9.7847e-01	8.5389e+01	6.4000e+01
3	6.4000e+01	1.6000e+01	1.2524e-01	1.7069e+01	1.6000e+01
4	1.6000e+01	4.0000e+00	7.8278e-03	4.0636e+00	4.0000e+00
5	4.0000e+00	1.0000e+00	1.9569e-03	1.0039e+00	1.0000e+00
6	1.0000e+00	2.5000e-01	4.8924e-04	2.5024e-01	2.5000e-01

Table 17: Highest and smallest photon energies $h\nu'$ from which photons are able to scatter into group g.

7.6.3 General case of the group transfer scattering cross section

First of all a general solution for arbitrary photon energies by help of numerical methods are presented. Secondly, for small photon energies an analytical solution is derived.

---

[7]This depends on the chosen energy groups.

7 MULTIGROUP APPROACH

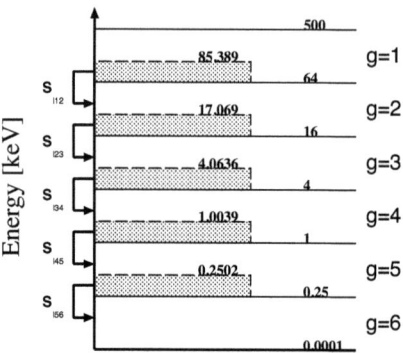

Figure 22: Graphical presentation of the inscattering contributions S_{lg-1g}.

The Compton differential cross section reads [Pomraning73]

$$\Sigma_s(\nu' \to \nu, \mu_0) = \frac{3}{16\pi} \Sigma_{Th} \frac{(1+\mu_0^2)}{(1+\gamma'(1-\mu_0))^2} \times$$
$$\times \left(1 + \frac{\gamma'(1-\mu_0)^2}{(1+\mu_0^2)(1+\gamma'(1-\mu_0))}\right) \delta\left(\nu - \nu'\left(\frac{1}{1+\gamma'(1-\mu_0)}\right)\right).$$

Factors of the same order of μ_0 are grouped together. Hence, the Compton differential cross section reads

$$\Sigma_s(\nu' \to \nu, \mu_0) = \frac{3}{16\pi} \Sigma_{Th} \frac{((1+\mu_0^2)(1+\gamma'(1-\mu_0)) + \gamma'(1-\mu_0)^2)}{(1+\gamma'(1-\mu_0))^3} \times$$
$$\times \delta\left(\nu - \frac{\nu'}{1+\gamma'(1-\mu_0)}\right) \quad (7.6.11)$$

which is transformed to

$$\Sigma_s(\nu' \to \nu, \mu_0) = \frac{3}{16\pi} \Sigma_{Th} \frac{(1+\gamma'+\gamma'^2) + (-\gamma'-2\gamma'^2)\mu_0 + (1+\gamma'+\gamma'^2)\mu_0^2}{(1+\gamma'(1-\mu_0))^3}$$
$$- \frac{\gamma'\mu_0^3}{(1+\gamma'(1-\mu_0))^3} \times \delta\left(\nu - \frac{\nu'}{1+\gamma'(1-\mu_0)}\right). \quad (7.6.12)$$

7.6 MOMENTS OF THE GROUP TRANSFER SCATTERING CROSS SECTION

(7.6.12) is inserted into (7.6.1). In this way one obtains a triple integral. Since we are asking for the scattering probability of photons with all possible frequencies ν' for all possible angles μ_0 the integral over $d\nu$ is evaluated by means of the δ-distribution. The resulting fraction on the right hand side is abbreviated by

$$\frac{P(\mu_0,\gamma')}{R(\mu_0,\gamma')} = \frac{(1+\gamma'+\gamma'^2)+(-\gamma'-2\gamma'^2)\mu_0+(1+\gamma'+\gamma'^2)\mu_0^2-\gamma'\mu_0^3}{(\gamma'^{-1}+1-\mu_0)^4}. \tag{7.6.13}$$

Hence, the moments of the Compton differential cross section read

$$S_{lg'g} = \frac{3}{8}\frac{\Sigma_{Th}}{B_{g'}}\frac{2(m_ec^2)^4}{c^2h^3}\int_{-1}^{1}d\mu_0\, P_l(\mu_0)\times$$
$$\times \int_{\gamma^{**}}^{\gamma^*}d\gamma'\frac{P(\mu_0,\gamma')}{R(\mu_0,\gamma')}\frac{1}{(\exp(\gamma'm_ec^2/\theta)-1)\gamma'} \tag{7.6.14}$$
$$= \frac{3}{8}\frac{\Sigma_{Th}}{B_{g'}}\frac{2(m_c^2)^4}{c^2h^3}\Gamma(\gamma^{**},\gamma^*) \tag{7.6.15}$$

where the results from the frequency shift formula (7.6.9) have been used. Here $\gamma^* = \gamma_g, \gamma^{**} = \gamma_{g+1}$ if $g' = g$ and $\gamma^* = \gamma_g/1-2\gamma_g, \gamma^{**} = \gamma_g$ if $g' = g-1$. $\Gamma(\gamma^{**},\gamma^*)$ is given by

$$\Gamma(\gamma^{**},\gamma^*) = \int_{\gamma^{**}}^{\gamma^*}d\gamma'\frac{1}{(\exp(\gamma'm_ec^2/\theta)-1)\gamma'}F_l(\gamma'), \tag{7.6.16}$$

$$F_l(\gamma') = \int_{-1}^{1}d\mu_0\frac{P(\mu_0,\gamma')}{R(\mu_0,\gamma')}P_l(\mu_0), = \sum_{i=1}^{4}F_l^{(i)}(\gamma') \tag{7.6.17}$$

where the integrals $F_l^{(i)}$ are defined by

$$F_l^{(1)}(\gamma') = \int_{-1}^{1}d\mu_0\frac{-\gamma'\mu_0^3}{(\gamma'^{-1}+1-\mu_0)^4}P_l(\mu_0) \tag{7.6.18}$$

$$F_l^{(2)}(\gamma') = \int_{-1}^{1}d\mu_0\frac{(1+\gamma'+\gamma'^2)\mu_0^2}{(\gamma'^{-1}+1-\mu_0)^4}P_l(\mu_0) \tag{7.6.19}$$

$$F_l^{(3)}(\gamma') = \int_{-1}^{1}d\mu_0\frac{(-\gamma'-2\gamma'^2)\mu_0}{(\gamma'^{-1}+1-\mu_0)^4}P_l(\mu_0) \tag{7.6.20}$$

$$F_l^{(4)}(\gamma') = \int_{-1}^{1}d\mu_0\frac{(1+\gamma'+\gamma'^2)}{(\gamma'^{-1}+1-\mu_0)^4}P_l(\mu_0). \tag{7.6.21}$$

The integrals (7.6.18-7.6.21) are analytically evaluable [Brytschkow92].

7 MULTIGROUP APPROACH

Example. Taking the Legendre polynomials of zeroth order $F_0^{(i)}$ reads

$$F_0^{(1)}(\gamma') = -\gamma'\left(\frac{3\tilde{a}}{1+\tilde{a}} - \frac{3\tilde{a}}{\tilde{a}-1} - \frac{3\tilde{a}^2}{2(1+\tilde{a})^2} + \frac{3\tilde{a}^2}{2(\tilde{a}-1)^2} + \frac{\tilde{a}^3}{3(1+\tilde{a})^3} - \frac{\tilde{a}^3}{3(\tilde{a}-1)^3} + \ln|1+\tilde{a}| - \ln|\tilde{a}-1|\right)$$

$$F_0^{(2)}(\gamma') = (1+\gamma'+\gamma'^2)\left(\frac{1}{\tilde{a}-1} - \frac{1}{1+\tilde{a}} + \frac{\tilde{a}}{(1+\tilde{a})^2} - \frac{\tilde{a}}{(\tilde{a}-1)^2} - \frac{\tilde{a}^2}{3(1+\tilde{a})^3} + \frac{\tilde{a}^2}{3(\tilde{a}-1)^3}\right)$$

$$F_0^{(3)}(\gamma') = (-\gamma' - 2\gamma'^2)\left(-\frac{1}{2(1+\tilde{a})^2} + \frac{1}{2(\tilde{a}-1)^2} + \frac{\tilde{a}}{3(1+\tilde{a})^3} - \frac{\tilde{a}}{3(\tilde{a}-1)^3}\right)$$

$$F_0^{(4)}(\gamma') = -\frac{1}{3}(1+\gamma'+\gamma'^2)\left(\frac{1}{(\tilde{a}+1)^3} - \frac{1}{(\tilde{a}-1)^3}\right)$$

$\tilde{a} = -1/\gamma' - 1$. The integral functions $F_l^{(i)}$ of higher order in l are evaluated in the same manner. $F_l^{(i)}$ evaluated in this way is inserted in (7.6.15). The integration over γ' in $S_{lg'g}$ is carried out numerically. The results are given in tables (18), (21) and (24) and are in perfect agreement with the ones obtained by Pritzker et al. [Pritzker76].

7.6.4 An asymptotic analytical solution

While $S_{lg'g}$ expressed by (7.6.14) is evaluable in a numerical way only, one can perform pure analytic calculations assuming the small photon energy limit $\gamma \ll 1$. In that case the differential scattering cross section $\Sigma_s(\nu' \to \nu, \mu_0)$ reads [Pomraning73]

$$\Sigma_s(\nu' \to \nu, \mu_0) = \frac{3}{16\pi}\Sigma_{Th}(1+\mu_0^2)\left[1 - 2\gamma(1-\mu_0) + \gamma^2\frac{(1-\mu_0)^2(4+3\mu_0^2)}{1+\mu_0^2}\right] \times \\ \times \delta\left(\nu' - \nu\left[1 - \gamma(1-\mu_0) + \gamma^2(1-\mu_0)^2\right]\right). \quad (7.6.22)$$

This term is inserted into (7.6.1). The resulting triple integral is evaluated over $d\nu$ by means of the δ-distribution. In this way one obtains

$$S_{lg'g} = \frac{3}{8}\frac{\Sigma_{Th}}{B_{g'}}\frac{2(m_ec^2)^4}{c^2h^3}\int_{-1}^{1}d\mu_0 P_l(\mu_0)\int_{\gamma^{**}}^{\gamma^*}d\gamma'\frac{\gamma'^3}{(\exp(\gamma'm_ec^2/\theta)-1)}\times \\ \times \left\{\left(1 - 3\gamma' + 7\gamma'^2 - 6\gamma'^3 + 4\gamma'^4\right) + (3\gamma' - 14\gamma'^2 + 18\gamma'^3 - 16\gamma'^4)\mu_0 + \right. \\ + \left(1 - 3\gamma' + 13\gamma'^2 - 23\gamma'^3 + 27\gamma'^4\right)\mu_0^2 + \left(3\gamma' - 12\gamma'^2 + 21\gamma'^3 - 28\gamma'^4\right)\mu_0^3 + \\ \left. + \left(6\gamma'^2 - 15\gamma'^3 + 22\gamma'^4\right)\mu_0^4 + \left(5\gamma'^3 - 12\gamma'^4\right)\mu_0^5 + 3\gamma'^4\mu_0^6\right\}. \quad (7.6.23)$$

7.6 Moments of the group transfer scattering cross section

In lowest Legendre order ($P_0(\mu_0) = 1$) the above expression integrated over μ_0 is

$$S_{0g'g} = \frac{3}{8}\frac{\Sigma_{Th}}{B_{g'}}\frac{2(m_ec^2)^4}{c^2h^3}\left\{\frac{8}{3}\int_{\gamma^{**}}^{\gamma^*}d\gamma'\frac{\gamma'^3}{(\exp(\gamma'm_ec^2/\theta)-1)}\right.$$
$$-8\int_{\gamma^{**}}^{\gamma^*}d\gamma'\frac{\gamma'^4}{(\exp(\gamma'm_ec^2/\theta)-1)} + \frac{376}{15}\int_{\gamma^{**}}^{\gamma^*}d\gamma'\frac{\gamma'^5}{(\exp(\gamma'm_ec^2/\theta)-1)} \quad (7.6.24)$$
$$\left.-\frac{100}{3}\int_{\gamma^{**}}^{\gamma^*}d\gamma'\frac{\gamma'^6}{(\exp(\gamma'm_ec^2/\theta)-1)} + \frac{1248}{35}\int_{\gamma^{**}}^{\gamma^*}d\gamma'\frac{\gamma'^7}{(\exp(\gamma'm_ec^2/\theta)-1)}\right\}$$

or written in terms as given in the appendix (A)

$$S_{0g'g} = \frac{3}{8}\frac{\Sigma_{Th}}{B_{g'}}\frac{2\theta^4}{c^2h^3}\left\{\frac{8}{3}\mathcal{I}_3(u^{**},u^*) - 8\alpha\mathcal{I}_4(u^{**},u^*) + \frac{376}{15}\alpha^2\mathcal{I}_5(u^{**},u^*)\right.$$
$$\left.-\frac{100}{3}\alpha^3\mathcal{I}_6(u^{**},u^*) + \frac{1248}{35}\alpha^4\mathcal{I}_7(u^{**},u^*)\right\} \quad (7.6.25)$$

where $\alpha = \theta/m_ec^2$ and $u^* = \gamma^*/\alpha$, $u^{**} = \gamma^{**}/\alpha$. The analytic solution is given while using (A.3.13) to (A.3.17). The result is presented in table (19). Proceeding with the next order group moment coefficient ($P_1(\mu_0)=\mu_0$) yields

$$S_{1g'g} = \frac{3}{8}\frac{\Sigma_{Th}}{B_{g'}}\frac{2(m_ec^2)^4}{c^2h^3}\left\{\frac{16}{5}\int_{\gamma^{**}}^{\gamma^*}d\gamma'\frac{\gamma'^4}{(\exp(\gamma'm_ec^2/\theta)-1)}\right.$$
$$-\frac{212}{15}\int_{\gamma^{**}}^{\gamma^*}d\gamma'\frac{\gamma'^5}{(\exp(\gamma'm_ec^2/\theta)-1)} + \frac{764}{35}\int_{\gamma^{**}}^{\gamma^*}d\gamma'\frac{\gamma'^6}{(\exp(\gamma'm_ec^2/\theta)-1)} \quad (7.6.26)$$
$$\left.-\frac{236}{5}\int_{\gamma^{**}}^{\gamma^*}d\gamma'\frac{\gamma'^7}{(\exp(\gamma'm_ec^2/\theta)-1)}\right\}$$

where its analytical solution is

$$S_{1g'g} = \frac{3}{8}\frac{\Sigma_{Th}}{B_{g'}}\frac{2\theta^4}{c^2h^3}\left\{\frac{16}{5}\alpha\mathcal{I}_4(u^{**},u^*) - \frac{212}{15}\alpha^2\mathcal{I}_5(u^{**},u^*) + \frac{764}{35}\alpha^3\mathcal{I}_6(u^{**},u^*)\right.$$
$$\left.-\frac{236}{5}\alpha^4\mathcal{I}_7(u^{**},u^*)\right\}. \quad (7.6.27)$$

The result is presented in table (22). Finally, the second order group moment ($P_2(\mu_0) = 0.5(-1 + 3\mu_0^2)$) is considered.

7 MULTIGROUP APPROACH

$$S_{2g'g} = -\frac{1}{2}S_{0g'g} + \frac{9}{16}\frac{\Sigma_{Th}}{B_{g'}}\frac{2(m_ec^2)^4}{c^2h^3}\left\{\frac{16}{15}\int_{\gamma^{**}}^{\gamma^*}d\gamma'\frac{\gamma'^3}{(\exp(\gamma'm_ec^2/\theta)-1)}\right.$$
$$-\frac{16}{5}\int_{\gamma^{**}}^{\gamma^*}d\gamma'\frac{\gamma'^4}{(\exp(\gamma'm_ec^2/\theta)-1)} + \frac{1216}{105}\int_{\gamma^{**}}^{\gamma^*}d\gamma'\frac{\gamma'^5}{(\exp(\gamma'm_ec^2/\theta)-1)} \quad (7.6.28)$$
$$\left.-\frac{612}{35}\int_{\gamma^{**}}^{\gamma^*}d\gamma'\frac{\gamma'^6}{(\exp(\gamma'm_ec^2/\theta)-1)} + \frac{2018}{105}\int_{\gamma^{**}}^{\gamma^*}d\gamma'\frac{\gamma'^7}{(\exp(\gamma'm_ec^2/\theta)-1)}\right\}$$

Solving the integrals leads to

$$S_{2g'g} = -\frac{1}{2}S_{0g'g} + \frac{9}{16}\frac{\Sigma_{Th}}{B_{g'}}\frac{2\theta^4}{c^2h^3}\left\{\frac{16}{15}\mathcal{I}_3(u^{**},u^*) - \frac{16}{5}\alpha\mathcal{I}_4(u^{**},u^*)\right.$$
$$\left.+\frac{1216}{105}\alpha^2\mathcal{I}_5(u^{**},u^*) - \frac{612}{35}\alpha^3\mathcal{I}_6(u^{**},u^*) + \frac{2018}{105}\alpha^4\mathcal{I}_7(u^{**},u^*)\right\}. \quad (7.6.29)$$

The result is presented in table (25). One can see by the tables (20), (23) and (26) the deviation from the numerically evaluated results is very small except for the highest energy group. Comparing to the published results from Pritzker one obtains the same results without much numerical effort.

7.6 MOMENTS OF THE GROUP TRANSFER SCATTERING CROSS SECTION

g',g/θ	1.00e-01	3.16e-01	1.00e+00	3.16e+00	1.00e+01
0,1	-	-	-	-	-
1,1	7.309e-01	7.302e-01	7.281e-01	7.206e-01	6.910e-01
1,2	7.309e-01	7.302e-01	7.281e-01	7.189e-01	5.329e-01
2,2	9.140e-01	9.129e-01	9.087e-01	8.906e-01	8.287e-01
2,3	9.140e-01	8.756e-01	5.366e-01	1.540e-01	2.059e-02
3,3	9.765e-01	9.749e-01	9.672e-01	9.457e-01	9.366e-01
3,4	4.354e-01	1.427e-01	2.812e-02	3.481e-03	1.176e-03
4,4	9.934e-01	9.906e-01	9.849e-01	9.831e-01	9.827e-01
4,5	2.821e-02	4.664e-03	6.242e-04	2.647e-04	2.073e-04
5,5	9.973e-01	9.960e-01	9.957e-01	9.956e-01	9.956e-01
5,6	7.448e-04	1.194e-04	6.118e-05	5.062e-05	4.783e-05
6,6	9.992e-01	9.991e-01	9.990e-01	9.990e-01	9.990e-01

Table 18: Zeroth order moments of the six-group Planck weighted Compton scattering transfer cross section. The results are given in units of the Thomson cross section $S_{lg'g}/\Sigma_{Th}$. The temperature θ is given in units of keV.

g',g/θ	1.00e-01	3.16e-01	1.00e+00	3.16e+00	1.00e+01
0,1	-	-	-	-	-
1,1	7.502e-01	7.497e-01	7.482e-01	7.430e-01	7.264e-01
1,2	7.502e-01	7.497e-01	7.482e-01	7.411e-01	5.533e-01
2,2	9.144e-01	9.133e-01	9.093e-01	8.916e-01	8.341e-01
2,3	9.144e-01	8.760e-01	5.369e-01	1.541e-01	2.060e-02
3,3	9.765e-01	9.749e-01	9.673e-01	9.459e-01	9.368e-01
3,4	4.354e-01	1.427e-01	2.812e-02	3.481e-03	1.176e-03
4,4	9.934e-01	9.906e-01	9.849e-01	9.831e-01	9.827e-01
4,5	2.821e-02	4.664e-03	6.242e-04	2.647e-04	2.073e-04
5,5	9.973e-01	9.960e-01	9.957e-01	9.956e-01	9.956e-01
5,6	7.448e-04	1.194e-04	6.118e-05	5.062e-05	4.783e-05
6,6	9.992e-01	9.991e-01	9.990e-01	9.990e-01	9.990e-01

Table 19: Zeroth order moment of the six-group Planck weighted Compton scattering transfer cross section in the case of small photon energies. The results are given in units of the Thomson cross section $S_{lg'g}/\Sigma_{Th}$. The approximation fails at high photon energy groups and high temperatures. In all other cases the results are in good agreement with those obtained by numerical integration. Refer to table (18) additionally.

g',g/θ	1.00e-01	3.16e-01	1.00e+00	3.16e+00	1.00e+01
0,1	-	-	-	-	-
1,1	2.651	2.673	2.761	3.106	5.120
1,2	2.646	2.673	2.761	3.097	3.813
2,2	0.047	0.049	0.057	0.119	0.652
2,3	0.047	0.048	0.050	0.050	0.051
3,3	0.001	0.001	0.003	0.014	0.021
3,4	0.001	0.001	0.001	0.001	0.001
4,4	0.000	0.000	0.000	0.000	0.000
4,5	0.000	0.000	0.000	0.000	0.000
5,5	0.000	0.000	0.000	0.000	0.000
5,6	0.000	0.000	0.000	0.000	0.000
6,6	0.000	0.000	0.000	0.000	0.000

Table 20: Accuracy of the multigroup Compton transfer cross section to zeroth order in the small photon energy limit. The deviation is given in percent comparing the general valid term (7.6.14) with small photon energy approximation (7.6.25). The results are in agreement to those achieved by the general valid expression (7.6.14) except for the highest energy groups.

7 MULTIGROUP APPROACH

g',g/θ	1.00e-01	3.16e-01	1.00e+00	3.16e+00	1.00e+01
0,1	-	-	-	-	-
1,1	9.291e-02	9.310e-02	9.370e-02	9.573e-02	1.032e-01
1,2	9.291e-02	9.310e-02	9.370e-02	9.544e-02	7.486e-02
2,2	3.307e-02	3.348e-02	3.499e-02	4.148e-02	6.237e-02
2,3	3.307e-02	3.203e-02	1.982e-02	5.711e-03	7.642e-04
3,3	9.313e-03	9.938e-03	1.290e-02	2.114e-02	2.459e-02
3,4	4.074e-03	1.336e-03	2.633e-04	3.260e-05	1.101e-05
4,4	2.632e-03	3.742e-03	5.995e-03	6.712e-03	6.873e-03
4,5	6.619e-05	1.094e-05	1.465e-06	6.212e-07	4.865e-07
5,5	1.072e-03	1.591e-03	1.726e-03	1.757e-03	1.766e-03
5,6	4.372e-07	7.010e-08	3.591e-08	2.971e-08	2.807e-08
6,6	4.010e-04	4.298e-04	4.368e-04	4.387e-04	4.393e-04

Table 21: First order moments of the six-group Planck weighted Compton scattering transfer cross section. The results are given in units of the Thomson cross section $S_{lg'g}/\Sigma_{Th}$. The temperature θ is given in units of keV.

g',g/θ	1.00e-01	3.16e-01	1.00e+00	3.16e+00	1.00e+01
0,1	-	-	-	-	-
1,1	7.891e-02	7.895e-02	7.909e-02	7.933e-02	7.566e-02
1,2	7.891e-02	7.895e-02	7.909e-02	7.914e-02	5.972e-02
2,2	3.279e-02	3.319e-02	3.465e-02	4.078e-02	5.861e-02
2,3	3.279e-02	3.175e-02	1.965e-02	5.660e-03	7.574e-04
3,3	9.308e-03	9.932e-03	1.289e-02	2.105e-02	2.446e-02
3,4	4.072e-03	1.335e-03	2.631e-04	3.258e-05	1.101e-05
4,4	2.632e-03	3.742e-03	5.993e-03	6.710e-03	6.871e-03
4,5	6.618e-05	1.094e-05	1.465e-06	6.212e-07	4.865e-07
5,5	1.072e-03	1.591e-03	1.726e-03	1.757e-03	1.766e-03
5,6	4.372e-07	7.010e-08	3.591e-08	2.971e-08	2.813e-08
6,6	4.010e-04	4.298e-04	4.368e-04	4.387e-04	4.393e-04

Table 22: First order moments of the six-group Planck weighted Compton scattering transfer cross section for the case of small photon energies. The results are given in units of the Thomson cross section $S_{lg'g}/\Sigma_{Th}$. The approximation fails at high photon energy groups and high temperatures. In all other cases the results are in very good agreement with those obtained by numerical integration. Refer to table (21) additionally.

g',g/θ	1.00e-01	3.16e-01	1.00e+00	3.16e+00	1.00e+01
0,1	-	-	-	-	-
1,1	15.072	15.192	15.584	17.129	26.703
1,2	15.076	15.192	15.584	17.084	20.218
2,2	0.846	0.872	0.980	1.702	6.019
2,3	0.846	0.867	0.885	0.892	0.894
3,3	0.054	0.063	0.134	0.412	0.521
3,4	0.052	0.052	0.052	0.052	0.052
4,4	0.004	0.011	0.028	0.033	0.034
4,5	0.003	0.003	0.003	0.003	0.003
5,5	0.001	0.002	0.002	0.002	0.002
5,6	0.000	0.000	0.000	0.000	0.207
6,6	0.000	0.000	0.000	0.000	0.000

Table 23: Accuracy of the multigroup Compton transfer cross section to first order in the small photon energy limit. The deviation in percent comparing the general valid term (7.6.14) with the small photon energy approximation (7.6.27). The results are in very good agreement to those achieved by the general valid expression (7.6.14) except for the highest energy groups.

7.6 MOMENTS OF THE GROUP TRANSFER SCATTERING CROSS SECTION

g',g/θ	1.00e-01	3.16e-01	1.00e+00	3.16e+00	1.00e+01
0,1	-	-	-	-	-
1,1	8.041e-02	8.039e-02	8.029e-02	7.997e-02	7.890e-02
1,2	8.041e-02	8.039e-02	8.029e-02	7.977e-02	5.987e-02
2,2	9.213e-02	9.204e-02	9.170e-02	9.029e-02	8.605e-02
2,3	9.213e-02	8.827e-02	5.411e-02	1.553e-02	2.076e-03
3,3	9.770e-02	9.755e-02	9.684e-02	9.489e-02	9.408e-02
3,4	4.356e-02	1.428e-02	2.814e-03	3.483e-04	1.177e-04
4,4	9.934e-02	9.907e-02	9.851e-02	9.834e-02	9.830e-02
4,5	2.821e-03	4.664e-04	6.242e-05	2.647e-05	2.073e-05
5,5	9.973e-02	9.960e-02	9.957e-02	9.956e-02	9.956e-02
5,6	7.448e-05	1.194e-05	6.118e-06	5.062e-06	4.783e-06
6,6	9.992e-02	9.991e-02	9.990e-02	9.990e-02	9.990e-02

Table 24: Second order moments of the six-group Planck weighted Compton scattering transfer cross section. The results are given in units of the Thomson cross section $S_{lg'g}/\Sigma_{Th}$. The temperature θ is given in units of keV.

g',g/θ	1.00e-01	3.16e-01	1.00e+00	3.16e+00	1.00e+01
0,1	-	-	-	-	-
1,1	8.486e-02	8.487e-02	8.490e-02	8.507e-02	8.665e-02
1,2	8.486e-02	8.487e-02	8.490e-02	8.484e-02	6.443e-02
2,2	9.224e-02	9.215e-02	9.184e-02	9.056e-02	8.735e-02
2,3	9.224e-02	8.838e-02	5.418e-02	1.555e-02	2.079e-03
3,3	9.770e-02	9.755e-02	9.684e-02	9.492e-02	9.413e-02
3,4	4.357e-02	1.428e-02	2.814e-03	3.483e-04	1.177e-04
4,4	9.934e-02	9.907e-02	9.852e-02	9.834e-02	9.830e-02
4,5	2.821e-03	4.664e-04	6.242e-05	2.647e-05	2.073e-05
5,5	9.973e-02	9.960e-02	9.957e-02	9.956e-02	9.956e-02
5,6	7.448e-05	1.194e-05	6.118e-06	5.062e-06	4.783e-06
6,6	9.992e-02	9.991e-02	9.990e-02	9.990e-02	9.990e-02

Table 25: Second order moments of the six-group Planck weighted Compton scattering transfer cross section in the case of small photon energies. The results are given in units of the Thomson cross section $S_{lg'g}/\Sigma_{Th}$. The approximation fails at high photon energy groups and high temperatures. In all other cases the results are in very good agreement with those obtained by numerical integration. Refer to table (24) additionally.

g',g/θ	1.00e-01	3.16e-01	1.00e+00	3.16e+00	1.00e+01
0,1	-	-	-	-	-
1,1	5.533	5.577	5.742	6.371	9.824
1,2	5.528	5.577	5.742	6.355	7.631
2,2	0.120	0.125	0.147	0.298	1.505
2,3	0.120	0.124	0.128	0.129	0.130
3,3	0.002	0.003	0.007	0.037	0.054
3,4	0.002	0.002	0.002	0.002	0.002
4,4	0.000	0.000	0.001	0.001	0.001
4,5	0.000	0.000	0.000	0.000	0.000
5,5	0.000	0.000	0.000	0.000	0.000
5,6	0.000	0.000	0.000	0.000	0.000
6,6	0.000	0.000	0.000	0.000	0.000

Table 26: Accuracy of the multigroup Compton transfer cross section of second order in small photon energy limit. The deviation is given in percent comparing the general valid term (7.6.14) with the small photon energy approximation (7.6.29). The results are in very good agreement with those achieved by the general valid expression (7.6.14) except for the highest energy groups.

7 Multigroup approach

8 Results

For the current investigation a one-dimensional spherical symmetric configuration consisting of an inner shell of D-T plasma subsequently followed by a shell of plutonium and an outer shell of explosive is considered. Refer to figure (23). Compared with the direct or indirect driven concept in ICF technology the explosive acts as driver. Contrary to ICF an additional shell consisting of plutonium has been added to the system surrounding the D-T plasma. That material is of relevance for depositing an additional energy contribution through fission processes.

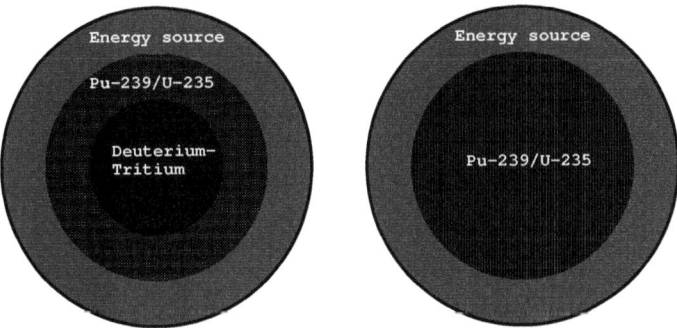

Figure 23: Layout of the physical model which is investigated. The physical processes depend on time. The fusion zone consists of a deuterium-tritium mixture while the second zone contains fissile material. In the present case these are the uranium isotop $^{235}_{92}$U and $^{239}_{94}$Pu. The inner cores is surrounded by an external energy source medium. In the current case this is some explosive. The main task of that external medium is the compression of the inner core. The configuration is surrounded by an iron shell.

In this section results of coupled fusion-fission calculations are presented. The influence of different approaches on the equation of states for D-T, the effectiveness of (radiation) heat conduction and the leverage of deposition of fusion energy on the hydrodynamic behaviour, fusion rates and thermonuclear burn-up will be presented.

In the present case the deposition of fusion energy is determined by the energy feed in of α- particles only. The energy of α-particles is deposited at the location where the fusion occurs. This mechanism is called α-heating. Energy contributions by neutrons have been

8 RESULTS

neglected. Calculations have been performed with and without α-heating. It is of interest how the α-heating influences the thermonuclear burn-up.

Calculations with a mixture of a few gram deuterium-tritium and plutonium of unit mass have been performed. The D-T plasma is split into 5 shells with increasing numbering from the innermost shell to the outermost. Besides studying the thermonuclear burn-up in each cell separately one defines the overall thermonuclear burn-up by

$$B(t_k) = \frac{1}{m} \sum_a m_a B_a(t_k), \qquad (8.0.30)$$

where a is the Lagrangian index of a zone, k the discrete time index, m_a is the mass and B_a the thermonuclear burn-up of deuterium-zone with Lagrangian index a, respectively.

The concept of ignition of the current configuration is comparable to those of an ICF-capsule [Pfalzner06]. The detonation of the explosive starts at time t = 0.0. Physically, the outer shells of the explosive blast off and the inner shells accelerate toward the center of the sphere. The acceleration is assumed to be spherical symmetric. The stability and symmetry of the implosion of the target is the most critical issue of the performance [Vehn97]. Instabilities and asymmetries leads to a decrease of the energy yield. The study of asymmetries requires at least 2-D codes. Due to the present 1-D configuration such effects are neglected. The explosive implodes toward the center and compresses the fissile and the D-T material. At appropriate condition the fissile material becomes overcritical and the nuclear energy release occurs. In the current model the nuclear chain reaction is initiated by a single neutron. The nuclear energy is released by an uncontrolled nuclear explosion. This heats up the fissile material significantly. Because of the strong compression the D-T core reaches densities of $10^3 - 10^5 \times$ D-T solid density and temperatures of 10 keV at the center. Fusion processes are ignitated at those conditions. Meyer-ter-Vehn [Vehn97] suggests similar densities for a thermonuclear burn-up of more than 10 percent. The incipient rise in pressure within D-T and the innermost Pu shells gives the main contribution for the rapid expansion of the configuration.

The nuclear energy yield of the current configuration is 9×10^6 MJ which is several orders larger comparing to published data of ICF facilities [Vehn97]. The fission rate rises to a peak of 10^9 TW. More powerful ICF facilities are planned or under construction [Bigot06].

8.1 The coupled system without α-heating

Zone/γ	no conduction		heat conduction		radh.* conduction	
	1.4	5/3	1.4	5/3	1.4	5/3
1	.93E+2	.95E+2	.94E+2	.95E+2	-	.93E+2
2	.11E-3	.86E+2	.51E+0	.89E+2	-	.93E+2
3	.10E-5	.58E+2	.11E-5	.68E+2	-	.93E+2
4	.11E-6	.34E+2	.15E-6	.45E+2	-	.93E+2
5	.13E-7	.22E+2	.16E-7	.30E+2	-	.93E+2

Table 27: Thermonuclear burn-up in a configuration without α-heating given in percent for a deuterium-tritium mixture of a few gram. *radiation heat conduction. γ is the adiabatic coefficient.

	no conduction		heat conduction		radh.* conduction	
Zone/γ	1.4	5/3	1.4	5/3	1.4	5/3
burn-up**	.16E+2	.56E+2	.16E+2	.63E+2	-	.93E+2

Table 28: Mass averaged thermonuclear burn-up in a configuration without α-heating given in percent in a deuterium-tritium mixture of a few gram. **Mass averaged thermonuclear burn-up given by equation (8.0.30). *radiation heat conduction. γ is the adiabatic coefficient.

Calculation: $\gamma = 1.4$, no conduction, no α-heating

In figure (25) the time histories of a calculation with an equation of state neglecting the equilibrium radiation term are shown. At the moment of maximum compression the temperature and density increase very sharply at time t ≈ 24.2085 μs because of the incoming shockwave from plutonium. The pressure is formed nearly homogeneously over all D-T zones. This is due to the equation of state for an ideal gas. The temperature within the 5 zones forms a very strong gradient from the innermost zone to the outermost zone. This is called the hot spot. The forming of the hot spot is shown in figure (27). Only the D-T plasma zones are shown. The zones of plutonium are skipped. Starting from $t = 24.190$ μs all zones are compressed by an incoming shockwave from plutonium. In this moment the temperature of the outermost D-T zone increases. The elevated temperature at snapshot $t = 24.190$ μs within the first D-T zone results from a preceding weaker hot spot. Normally, the temperature is at lowest here. Proceeding in time the zones are heated up and

8 RESULTS

forms a hot spot with highest temperature in the centre. The temperature gradient is immediately dissipated by conduction effects but these effects have been neglected for this calculation. Due to the low temperatures in zones 2 to 5 the thermonuclear burn-up within these zones is neglectable. The D-T within the first cell is burned-up up to 93 percent. See table (27). The overall thermonuclear burn-up for the performed calculation is 15.7 percent. See table (28). The wave structure in all quantities results from partial reflections of the outgoing shockwave in D-T on the material border to plutonium. Because of the different densities of D-T and Pu parts of this wave are rejected back to the D-T plasma, whereas other parts penetrate the fissile material. The first negative hump of the velocity shows the input of a strong shockwave from plutonium. The vector of velocity is directed to the centre of core. At the (t $\approx$ 24.2085 μs) the velocity vector immediately changes its direction and the D-T plasma tries to expand. The first positive hump indicates the first reflection on the fissile material. Figure (26) shows snapshots of the behaviour of the Rosseland and Planck mean free path and the expansion radius of D-T and plutonium depending on time. The Lagrangian coordinate indicates the zone number. The average mean free paths and the expansion radius of D-T are shown in red colour. Except for the zones 2-5 of the D-T plasma the Rosseland mean free path are almost always larger than the expansion radius. That means, the system is optical thin.

Calculation: $\gamma = 5/3$, no conduction, no α-heating

A similar situation is obtained by taking an equation of state of an ideal gas with an equilibrium radiation term into account. The results of such a calculation are shown in figure (28). Comparing the temperature behaviour in figures (25) and (28) one recognises that the temperature is approximately one order higher within all zones. This is due to the radiation term in the equation of state. The higher temperature results in a higher rate of fusion and an increasing thermonuclear burn-up. The pressure is homogenous within all zones and behaves similar as in the prior case. Therefore and because of an ideal gas approach the increase of temperature results in a decrease in density. As a result of high temperatures a significant amount of energy is produced by radiation.

8.1 THE COUPLED SYSTEM WITHOUT α-HEATING

Calculation: $\gamma = 1.4$, electronic heat conduction, no α-heating

Figure (29) presents the results of a calculation taking pure heat conduction into account. The calculation has been performed considering an ideal gas equation of state with an adiabatic coefficient $\gamma = 1.4$ without equilibrium radiation term. The time scale is expanded to show the conduction behaviour. Starting from t $\approx$ 24.325 μs the temperature of zone 2 increases significantly by heat energy transport. At later times t $\approx$ 25.000 μs the temperature of zone 3 starts to increase by heat energy transport too. The beginning of increasing of temperature by conduction is marked by arrows.

Calculation: $\gamma = 5/3$, electronic heat conduction, no α-heating

The results of a calculation using an equation of state with an adiabatic coefficient $\gamma = 5/3$ and taking pure heat conduction contribution into account are not presented, because of its very small effect. The effect is shown in figure (24) by plotting the temperature of the innermost and second zone obtained by calculations neglecting as well as involving conduction phenomena. Comparing to the calculation disregarding the equilibrium radiation term in the equation of state one recognises, that in the current configuration heat conduction plays a role at temperatures below one keV only.

Calculation: $\gamma = 5/3$, radiation heat conduction, no α-heating

The radiation heat conduction plays an important role. Figure (30) displays the results of a calculation where $\gamma = 5/3$ and the equilibrium radiation term are involved. The time scale is reduced. The temperature of all zones is immediately heated up by radiation heat conduction. Therefore the D-T is burned up to 93 percent within all D-T zones.

8 Results

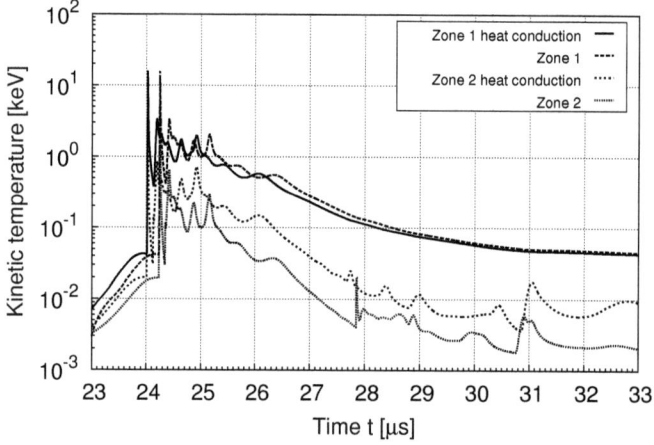

Figure 24: Comparison of calculations with and without pure heat conduction by using an equation of state with $\gamma = 5/3$ and taking the equilibrium radiation contribution into account. Only the innermost and second zone are shown. The time shift of $\Delta t \approx 0.3\ \mu s$ results from an increase of iteration loops followed by an increase of numerical errors in conduction calculations. The pure heat conduction is of small effect.

8.2 The coupled system including α-heating

Zone/γ	no conduction		heat conduction		radh.* conduction	
	1.4	5/3	1.4	5/3	1.4	5/3
1	.89E+2	.82E+2	-	.91E+2	-	-
2	.98E+2	.89E+2	-	.97E+2	-	-
3	.99E+2	.91E+2	-	.97E+2	-	-
4	.99E+2	.92E+2	-	.98E+2	-	-
5	.99E+2	.91E+2	-	.98E+2	-	-

Table 29: Thermonuclear burn-up in a configuration taking α-heating into account given in percent for a D-T plasma of a few gram. Because of the almost complete thermonuclear burn-up by α-heating the calculations taking radiation heat conduction into account have been skipped. *radiation heat conduction. γ is the adiabatic coefficient.

Zone/γ	no conduction		heat conduction		radh.* conduction	
	1.4	5/3	1.4	5/3	1.4	5/3
burn-up**	.97E+2	.89E+2	-	.96E+2	-	-

Table 30: Mass averaged thermonuclear burn-up in a configuration without α-heating given in percent in a deuterium-tritium mixture of a few gram. **Mass averaged thermonuclear burn-up given by equation (8.0.30). *radiation heat conduction. γ is the adiabatic coefficient.

The situation significantly changes when fusion energy is deposited by α-particles in the system. The same configuration as in case without α-heating has been used. As one can see in equation (3.5.6) the approximation of the fusion cross section is strongly influenced by the temperature. Slight changes in temperature result in strong changes of the fusion cross section. The temperature is determined by the specific heat capacity c_V. In the current configuration c_V is constant. The time scale is reduced to $\Delta t \approx 0.1$ μs for all calculations. Other time periods provide no further physical important details.

Calculation: $\gamma = 1.4$, no conduction, α-heating

The first calculation including α-heating has been performed using an ideal gas equation of state with $\gamma = 1.4$. The results are shown in figure (31). The temperatures and pressures are 2 orders of magnitude higher as compared to the same configuration without α-heating.

8 RESULTS

This raise in temperature seems to be unphysical. Because of the constant specific heat the temperature is related to the internal energy. Extreme temperatures are obtained by performing calculations using an ideal gas equation of state and the choice of $\gamma = 5/3$. The deposition of fusion energy by α-particles leads to a coupling between the increase of temperature and enhancement of fusion rates and thermonuclear burn-up. Therefore D-T plasma is almost completely burned within all zones. One can expect, that energy transport effects do not have any significant influence on the increase of burn-up. The burn-up of the first D-T plasma zone is slightly smaller as compared to the case without α-heating. This is due to the lower density of the first zone. Due to the high temperatures and low densities of all zones the system is transparent for radiation starting from $t \approx 24.21\ \mu s$.

Calculation: $\gamma = 1.4$, no conduction, α-heating

Similar to the case without α-heating a large amount of energy is present by heat radiation. Taking an ideal gas equation including the equilibrium radiation term into account should lower the temperature within all zones. This is shown in figure (32). The temperature is evaluated by an internal iteration procedure and does not depend on the choice of c_V.

Calculation: $\gamma = 5/3$, electronic heat conduction, α-heating

The results of a calculation performed with an ideal gas equation using $\gamma = 5/3$ and equilibrium radiation energy term and electronic heat conduction are shown in figure (33). Due to the weak temperature gradient the lowering of temperature of the innermost zone and the temperature of the second zone respectively are not seen clearly. The influence of conduction is of percent. Refer to table (30). Referring to table (29) the D-T plasma is burned almost completely.

8.3 Summary

In this thesis the thermonuclear burn-up in D-T plasma ignited by an uncontrolled thermonuclear explosion has been studied. In conclusion one can say that the D-T plasma is mainly burned-up in the moment of highest compression and temperature due to the incoming shockwave from Pu. Further important influences to the burn-up are the choices of the equation of state, the conduction mechanism and the contribution of the fusion energy. On the one hand the D-T plasma is taken into account as ideal gas and on the other hand it is investigated by an ideal gas equation extended for an equilibrium radiation term. The latter case is referred as advanced equation of state.

The amount of thermonuclear burn-up is calculated for the case where all physical fusion quantities are calculated but the fusion energy is assumed to be small and has been neglected therefore. This is an approximation of systems where the produced fusion rates and fusion energies do not lead to a significant enhancement of the temperature within the D-T plasma. For example this happens in strongly asymmetric compressions.

Independently from the equation of the D-T plasma the temperature within all zones forms a very strong gradient. The highest temperature appears in the innermost D-T plasma zone. Due to the strong dependency of the D-T fusion cross section on temperature the thermonuclear burn-up is very high within this zone. The results of simulations estimating the equation of state of a D-T plasma by an ideal gas equation and by an ideal gas equation taking the equilibrium radiation term into account show, that the equilibrium radiation term strongly influences the behaviour of the D-T material. This term reduces the temperature of the first D-T material zone and elevates the temperatures in all other D-T material zones significantly. Hence, the simple equation of state ansatz by an ideal gas leads to a large thermonuclear burn-up in the first D-T zone only. The advanced equation of state results in a high burn-up within all D-T plasma zones.

The burn-up is additionally influenced by conduction. Pure heat conduction plays a role at low temperatures only. This effect influences the system in case of taking an ideal gas equation as equation of state for the D-T plasma only. The D-T plasma is immediately burned when radiation heat conduction is taken into account. Radiation heat conduction strongly influences and enhances the temperature and burn-up of all zones.

The situation is immediately changed, when the kinetic energy of α- particles is considered

8 RESULTS

as external energy source of the coupled system. This additional energy leads to an increase in temperature within all D-T plasma zones. The D-T material is immediately burned-up to 90 percent approximately. The choice of a simple ideal gas as equation of state leads to unphysical temperatures of several hundred keV in the innermost D-T plasma zone. Due to the high temperature a large amount of energy is present by radiation. Hence, the usage of the advanced equation of state reduced the temperatures down to 30 keV approximately. Due to the high burn-up the conduction mechanism plays a minor role only.

Starting from the nuclear energy release the temperature increases and the innermost D-T plasma zone and the plutonium zones becomes optical thin. The fissile material as like as the D-T plasma becomes transparent for radiation. Hence, solving the problem of radiative transfer by a diffusion approach is not valid. One has to use a flux limited approach or find a solution for the general radiative transfer equation.

As mentioned earlier the compression of targets by laser and particle beams or explosive is strongly accompanied by asymmetrical and instability[8] processes. These leakage processes reduce the fusion energy output. Therefore, the conduction becomes important to the thermonuclear burn-up, when the thermal energy by fusion is neglectable. As compared to the compression of the fuel by lasers or X-rays the compression by shockwaves as a result of uncontrolled nuclear chain reactions is much more powerful.

---

[8]Known as Rayleigh – Taylor instability [Pfalzner06].

8.3 SUMMARY

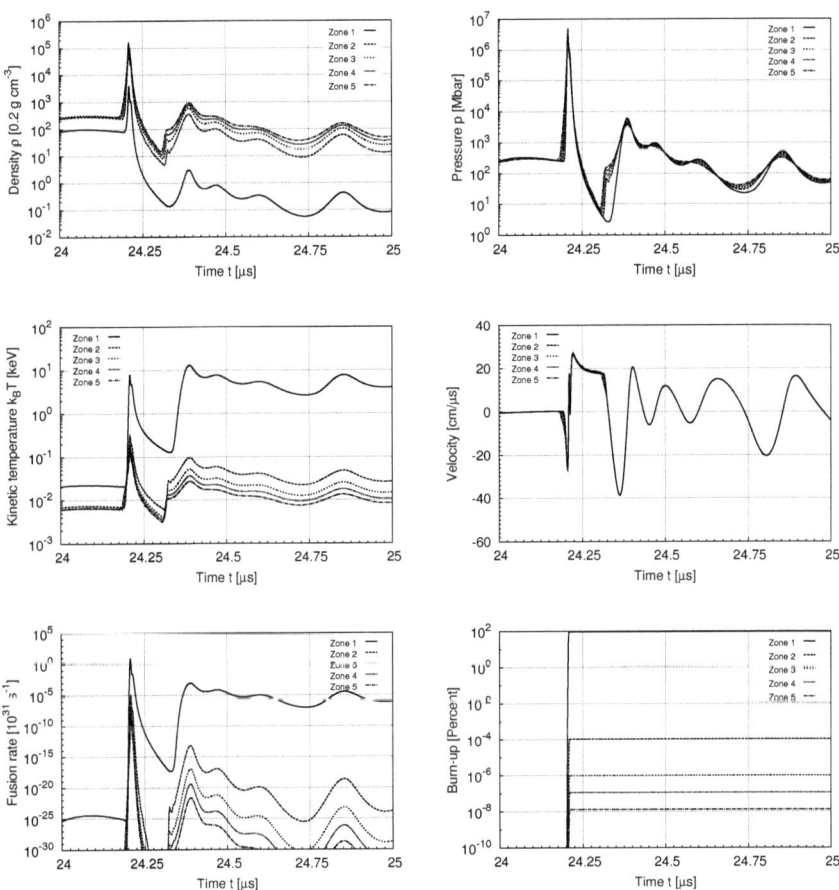

Figure 25: Results of a calculation where the D-T plasma is described by an equation of state of an ideal gas with $\gamma = 1.4$. No fusion energy is deposited in the physical system. Conduction calculations are switched off.

8 RESULTS

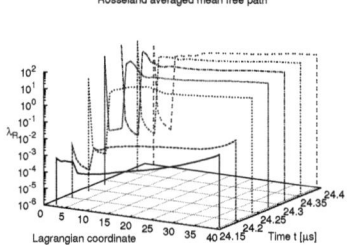
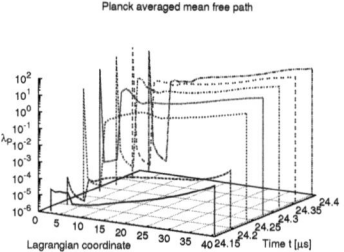

Figure 26: Snapshots of the Rosseland and Planck mean free path and of the outer radii of D-T and plutonium. The Lagrangian coordinate means the spherical zone index of the configuration. The Rosseland and Planck mean free path are given in cm.

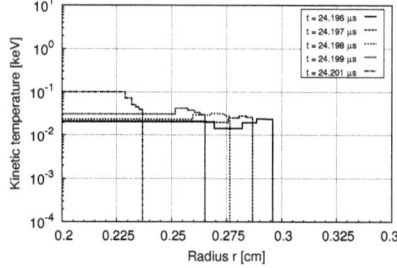
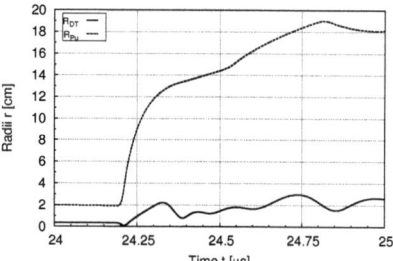

Figure 27: Forming of the hot spot without conduction effects. An ideal gas equation of state with $\gamma = 1.4$ without equilibrium radiation contributions has been used. The elevated temperature at t = 24.196 μs within the first D-T zone results from a preceding weaker hot spot.

8.3 Summary

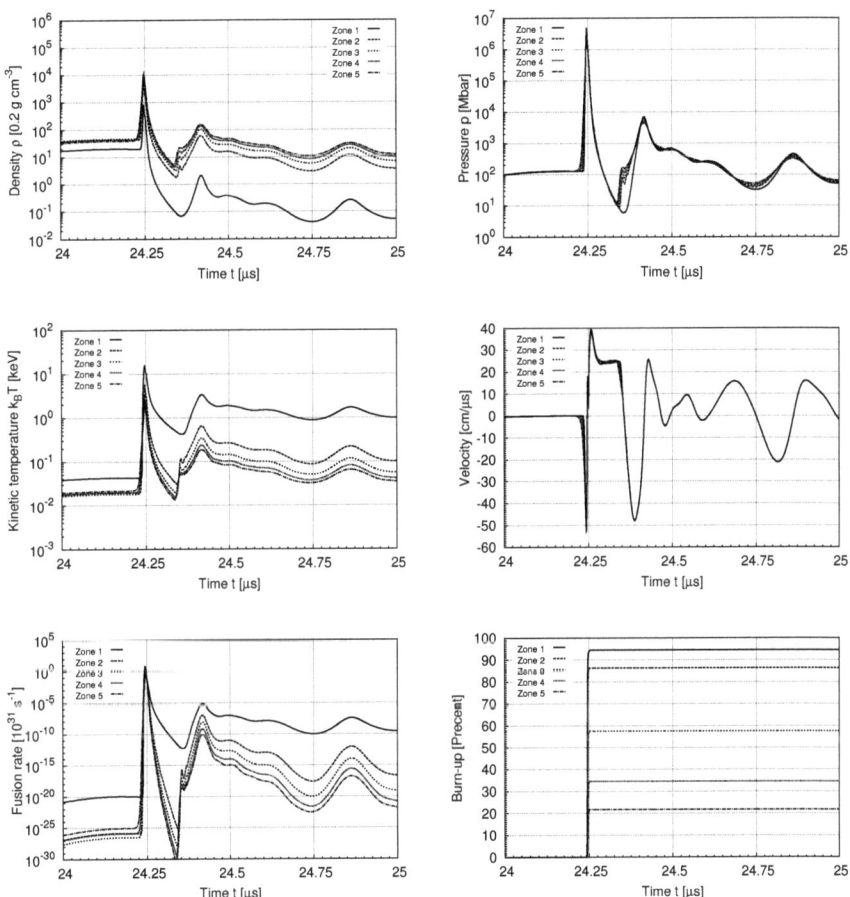

Figure 28: Results of a calculation where the D-T plasma is described by an equation of state of an ideal gas choosing $\gamma = 5/3$ taking equilibrium radiation contributions into account. No fusion energy is deposited in the system. All conduction calculations are switched off.

8 RESULTS

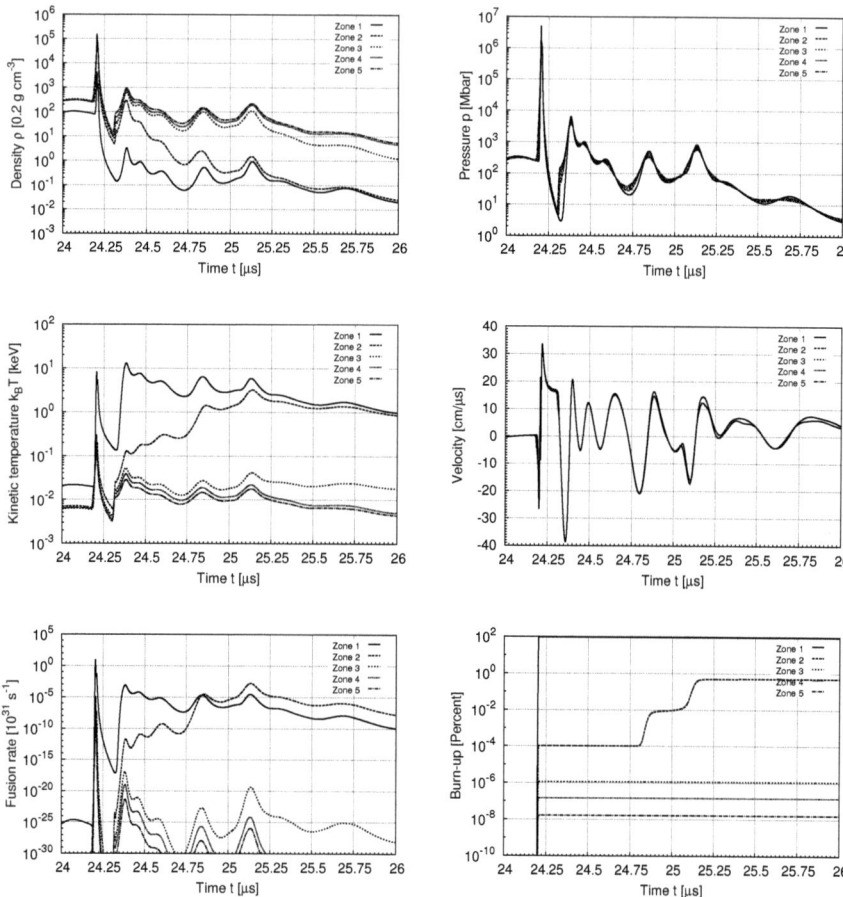

Figure 29: Results of a calculation where the D-T plasma is described by an equation of state of an ideal gas with $\gamma = 1.4$. No fusion energy is deposited in the physical system. Conduction has been considered as pure heat conduction in D-T and plutonium by the ansatz of Spitzer. Refer to equation (6.1.2).

8.3 SUMMARY

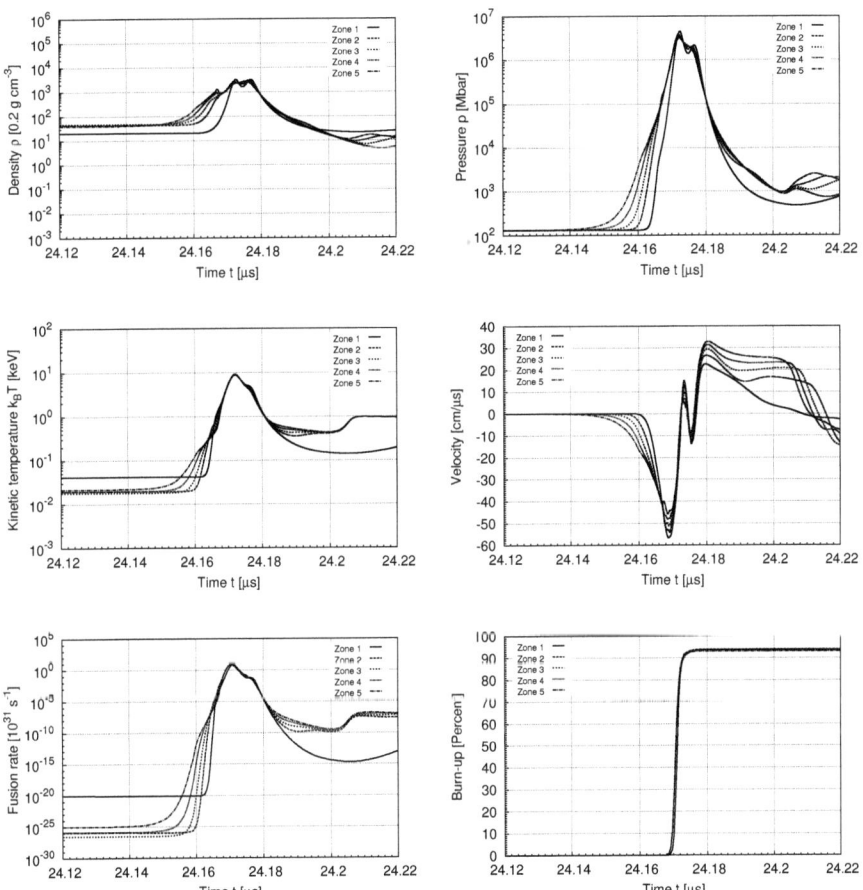

Figure 30: Results of a calculation where the D-T plasma is described by an equation of state of an ideal gas with $\gamma = 5/3$ taking equilibrium radiation contributions into account. No fusion energy is deposited in the physical system. Conduction has been considered as heat and radiation conduction in D-T and plutonium. Refer to equations (6.2.15) and (6.1.2).

8 RESULTS

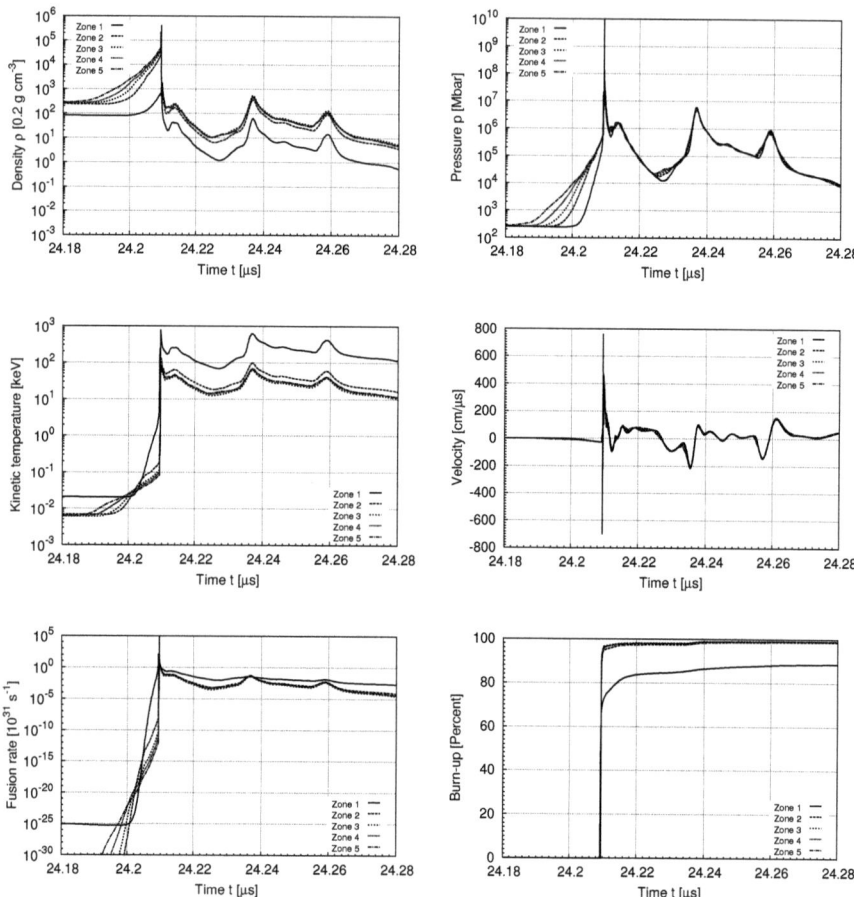

Figure 31: Results of a calculation where the D-T plasma is described by an ideal gas with $\gamma = 1.4$ neglecting radiation energy contributions. Fusion energy is deposited in the system. Conduction effects are not considered.

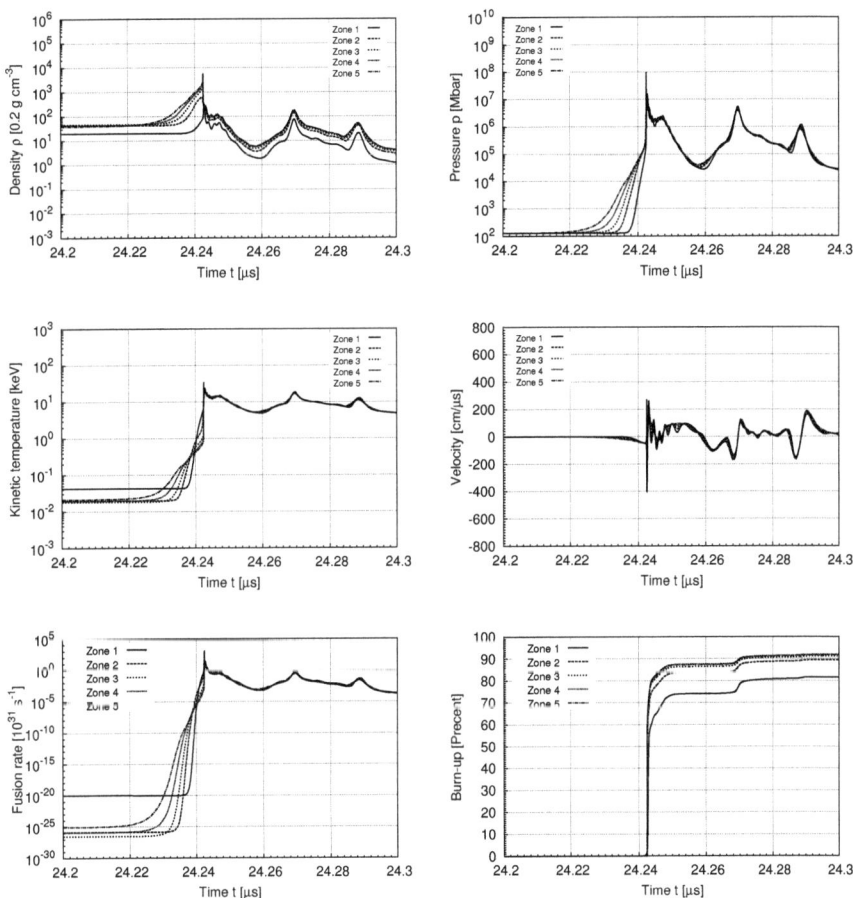

Figure 32: Results of a calculation where the D-T plasma is described by an ideal gas choosing $\gamma = 5/3$ with radiation contribution. Fusion energy is deposited in the system. Conduction effects are not considered.

8 RESULTS

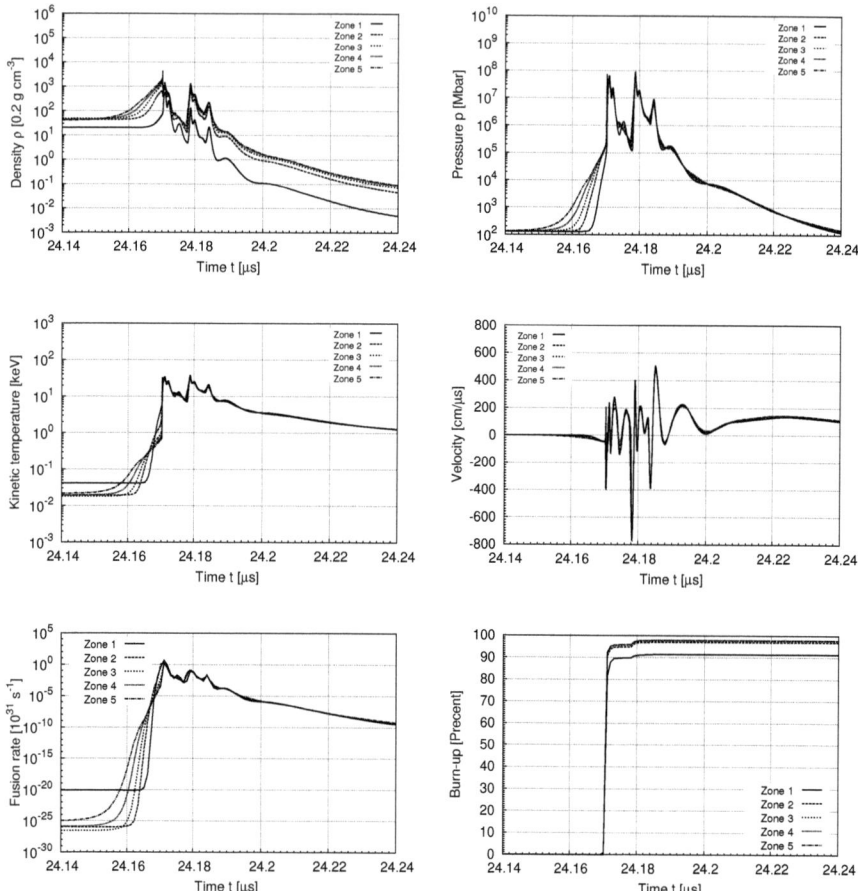

Figure 33: Results of a calculations where the D-T plasma is described by an ideal gas choosing $\gamma = 5/3$ with radiation contribution. Fusion energy is deposited in the system. Conduction has been considered as pure heat conduction in D-T and plutonium by the ansatz of Spitzer. See equation (6.1.2).

9 Program System

The presented results have been generated using a coupled program system consisting of the pure hydrodynamic code STEALTH , the neutron transport code MCNP and the external library enrico for generating the radiation mean free paths. MCNP is based on an Eulerian coordinate system whereas STEALTH uses the lagrangian coordinate presentation. The hydrodynamic equations are solved by an explicit finite fully difference scheme. STEALTH has significantly been extended by fusion rate and energy calculations, coupling routines to MCNP , fission energy calculations and (radiation) heat conduction routines. Neutron cross sections from the internal ENDF library of MCNP have been used. It should be noted, that these cross sections do not depend on temperature. For the purpose of validation of the results with respect to a former program system, the neutron cross section are grouped in six parts depending on neutron energy. Neutrons with energies below 50 eV and above 10 MeV are skipped. Due to the deposition of fusion neutrons in the system, this group configuration has to be extended up to 14 MeV in future processes.

The internal time step behaviour of STEALTH is determined by the sound speed squared (3.4.29) and the thermal diffusivity (6.2.16). The numerical stability of STEALTH is assured by the criterion of Courant. Unfortunately, this is computational very cost intensive for the case of radiation heat conduction calculations.

A specific model is setup by a zone configuration with the possibility of different cell spacing. Every zone contains a material configured by an equation of state, reference density, initial temperature and boundary conditions.

At time $t = 0.0$ the system is initiated. During the first few μs purely hydrodynamic calculations are performed. Approximately 1 μs before the system becomes overcritical coupled calculations between STEALTH and MCNP are started. Within this time period k_{eff} is calculated by MCNP . All other neutron physical quantities are ignored. k_{eff} acts as an indicator when the system becomes overcritical, that means $k_{eff} \geq 1$. When this condition is fulfilled, the criticality calculations will be stopped. In critical and overcritical regions and later when the system becomes undercritical again MCNP gives neutron fluxes, fission rates and neutron densities back. An external routine calculates the nuclear fission energy by taking these values.

STEALTH and MCNP are weakly coupled. Refer to figure (34). That means after a loop of

9 PROGRAM SYSTEM

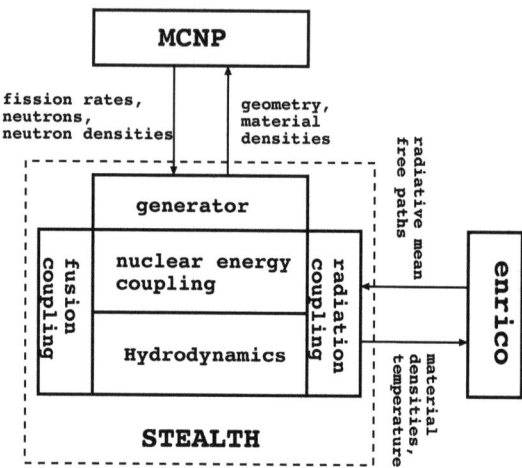

Figure 34: Schematic view of the program system used for the presented calculations. STEALTH consists of different parts responsible for calculating and deposition of the fusion energies, fusion rates, deposition of fission energy and heat fluxes. The radiation mean free paths calculated by enrico are based on the material densities and material temperatures obtained from STEALTH.

STEALTH is finished an input of MCNP taking the actual values of geometry and material densities is generated by an external routine and MCNP is executed. When finishing a MCNP loop the data are analysed separately and given back to STEALTH. The part enrico is linked to STEALTH and executed for every material zone within each time step of STEALTH.

10 Conclusion and Outlook

In this thesis the results from a coupled nuclear fission-fusion configuration taking conduction into account are presented. For that the hydrodynamical equations have been expanded in such a way that nuclear and fusion energy have been considered and implemented as external energy source terms. The partial (integro-) differential equations are solved numerically by setting up a coupling between the hydrodynamic program STEALTH and the neutron transport program MCNP. Moreover the equation of state for the deuterium-tritium plasma has been extended with respect to the appearance of thermal radiation at high temperatures. This additional term significantly reduces the temperature in the D-T plasma within all cases. The equation of state for fissile material has been given by a 3-term series expansion in temperature. The external energy source is described by an ideal gas equation.

Based on the knowledge of material densities and temperatures the averaged radiation mean free paths valid in the optical thick and thin limit are calculated. The Rosseland mean free path as valid for a optical thick regime has been used while deriving the radiation heat conductivity coefficient. Radiation conduction significantly reduce the complexity of radition transport. The conduction ansatz by a law of Fourier is valid for small gradients in temperature only. Therefore, conduction calculations have been performed by an ansatz of flux limiters for both the electronic and radiation heat conduction. It is shown that conduction plays an important role in burning the D-T plasma.

It has been shown that the innermost D-T zone is almost optical thin whereas all other zones nearly achieves the optical thick limit. Numerically, this depends on the number of zones.

Beside numerical studies the equation of radiative transfer has been studied analytically in a multigroup approximation. An approximation for the absorption, scattering and in-scattering cross section are derived. The results are in very good agreement with the ones studied numerically in literature [Pritzker76].

One of the next steps is the feed in of fusion neutrons to the system. It is to be expected, that fusion neutrons increase significantly the nuclear burn-up of the fission material. In that case the enhancement in released fission energy leads to an enhancement of the fusion processes itself. The interaction between fusion and fission will be stopped immediately

10 Conclusion and Outlook

when fusion conditions are not further fulfilled.

A Appendix

A.1 Units

The following constants are taken from [Zel'dovich66, Allen64].

$$
\begin{aligned}
e &= 4.803 \times 10^{-10} \text{ esu} = 4.803 \times 10^{-10} \text{ cm}\sqrt{\text{g cm/s}^2} \\
m_e &= 9.108 \times 10^{-28} \text{ g} \\
c &= 2.998 \times 10^{10} \text{ cm/s} = 2.998 \times 10^5 \text{ cm}/\mu\text{s} \\
h &= 4.136 \times 10^{-15} \text{ eV} \times \text{s} = 6.62607 \times 10^{-27} \text{ erg} \times \text{s} \\
A &= 6.022 \times 10^{23} \text{ mol}^{-1} \\
k_B &= 8.617 \times 10^{-5} \text{ eV/K} = 1.380 \times 10^{-16} \text{ erg/K} \\
\sigma_{SB} &= \frac{2\pi^5 k_B^4}{15 h^3 c^2} = 5.67 \times 10^{-11} \text{erg/cm}^2/\mu\text{s}/\text{K}^4 \\
a_{rad} &= \frac{8\pi^5 k_B^4}{15 h^3 c^3} = \frac{4}{c}\sigma_{SB} = 7.566 \times 10^{-15} \text{erg/cm}^3/\text{K}^4 \\
\epsilon_0 &= 8.854 \times 10^{-12} \text{A}^2\text{s}^4/\text{kg}/\text{m}^3
\end{aligned}
$$

A.2 Zeta-function

The zeta-function solves the integral

$$I = \frac{1}{k!} \int_0^\infty du \, \frac{u^{k-1}}{\exp(u) - 1} \tag{A.2.1}$$

and is given by [Smirnow95]

$$\zeta(k) = \sum_{n=0}^\infty \frac{1}{(n+1)^k}. \tag{A.2.2}$$

A.3 Integration of Planck's function

The source function in (local) thermal equilibrium is given by the Planck function

$$B_\nu(\theta) = \frac{2h\nu^3}{c^2} \left(\exp\left(h\nu/\theta\right) - 1\right)^{-1}, \tag{A.3.1}$$

where c is the speed of light, h is the Planck constant, ν the photon frequency and θ the temperature. Of widely use within this thesis are the integrals

A Appendix

$$I_1 = \int_0^\infty d\nu \, B_\nu(\theta) \tag{A.3.2}$$

$$I_2 = \int_0^\infty d\nu \, \frac{\partial B_\nu(\theta)}{\partial \theta}. \tag{A.3.3}$$

Those integrals are solved in the following way

$$I_1 = \int_0^\infty d\nu \, \frac{1}{h^2} \frac{2h^3\nu^3}{c^2} \left(\exp(h\nu/\theta) - 1\right)^{-1}. \tag{A.3.4}$$

By defining $u = h\nu/\theta$ one has

$$I_1 = \frac{2}{c^2} \frac{\theta^4}{h^3} \int_0^\infty du \, u^3 \left(\exp(u) - 1\right)^{-1} = \frac{2}{c^2} \frac{\theta^4}{h^3} \int_0^\infty du \, u^3 \sum_{n=0}^\infty \exp(-(n+1)u). \tag{A.3.5}$$

With $y = (n+1)u$ the solution is[9]

$$I_1 = \frac{2}{c^2} \frac{\theta^4}{h^3} \sum_{n=0}^\infty \frac{1}{(n+1)^4} \int_0^\infty dy \, y^3 \exp(-y) = \frac{2\pi^4}{15} \frac{\theta^4}{c^2 h^3} = \frac{ac}{4\pi} T^4. \tag{A.3.6}$$

The relation [Brytschkow92]

$$\int_0^\infty dy \, y^n \exp(-y) = n! \tag{A.3.7}$$

has been used while deriving the above result. By help of (A.3.6) the integral I_2 is integrated in a similar way

$$I_2 = \frac{\partial}{\partial \theta} \int_0^\infty d\nu \, B_\nu(\theta) = \frac{8\pi^4}{15} \frac{\theta^3}{c^2 h^3} = \frac{ac}{\pi} T^3. \tag{A.3.8}$$

Some more effort is required while solving integrals of type (A.2.1) in a given arbitrary real range [a,b]

$$\mathcal{I}_k(a,b) = \int_a^b du \, \frac{u^k}{\exp(u) - 1} = \sum_{n=0}^\infty \frac{1}{(n+1)^{k+1}} \int_{(n+1)a}^{(n+1)b} dy \, y^k \exp(-y). \tag{A.3.9}$$

Evaluation of the last integral leads to the final result

$$\mathcal{I}_k(a,b) = \sum_{n=0}^\infty \frac{1}{(n+1)^{k+1}} \left[\exp(-\tilde{a}) \left(\sum_{i=0}^k \frac{k!}{i!} \tilde{a}^i \right) - \exp(-\tilde{b}) \left(\sum_{i=0}^k \frac{k!}{i!} \tilde{b}^i \right) \right], \tag{A.3.10}$$

---

[9] The equilibrium radiation energy is $U = \frac{4\pi}{c} I_1$.

A.3 INTEGRATION OF PLANCK'S FUNCTION

where $\tilde{a} = (n+1)a$ and $\tilde{b} = (n+1)b$. $\mathcal{I}_k(a,b)$ is of widely use within the investigation of multigroup radiation cross sections.

Example. For $k = 1, \ldots, 7$ $\mathcal{I}_k(a,b)$ reads

$$\mathcal{I}_1(a,b) = \int_a^b du \frac{u}{\exp(u) - 1} = \sum_{n=0}^{\infty} \frac{1}{(n+1)^2} \left[\exp(-\tilde{a})(\tilde{a} + 1) - \exp(-\tilde{b})(\tilde{b} + 1) \right] \tag{A.3.11}$$

$$\mathcal{I}_2(a,b) = \int_a^b du \frac{u^2}{\exp(u) - 1} = \sum_{n=0}^{\infty} \frac{1}{(n+1)^3} \left[\exp(-\tilde{a})(\tilde{a}^2 + 2\tilde{a} + 2) - \exp(-\tilde{b})(\tilde{b}^2 + 2\tilde{b} + 2) \right] \tag{A.3.12}$$

$$\mathcal{I}_3(a,b) = \int_a^b du \frac{u^3}{\exp(u) - 1} = \sum_{n=0}^{\infty} \frac{1}{(n+1)^4} \left[\exp(-\tilde{a})(\tilde{a}^3 + 3\tilde{a}^2 + 6\tilde{a} + 6) - \exp(-\tilde{b})(\tilde{b}^3 + 3\tilde{b}^2 + 6\tilde{b} + 6) \right] \tag{A.3.13}$$

$$\mathcal{I}_4(a,b) = \int_a^b du \frac{u^4}{\exp(u) - 1} = \sum_{n=0}^{\infty} \frac{1}{(n+1)^5} \left[\exp(-\tilde{a})(\tilde{a}^4 + 4\tilde{a}^3 + 12\tilde{a}^2 + 24\tilde{a} + 24) - \exp(-\tilde{b})(\tilde{b}^4 + 4\tilde{b}^3 + 12\tilde{b}^2 + 24\tilde{b} + 24) \right] \tag{A.3.14}$$

$$\mathcal{I}_5(a,b) = \int_a^b du \frac{u^5}{\exp(u) - 1} = \sum_{n=0}^{\infty} \frac{1}{(n+1)^6} \left[\exp(-\tilde{a})(\tilde{a}^5 + 5\tilde{a}^4 + 20\tilde{a}^3 + 60\tilde{a}^2 + 120\tilde{a} + 120) - \exp(-\tilde{b})(\tilde{b}^5 + 5\tilde{b}^4 + 20\tilde{b}^3 + 60\tilde{b}^2 + 120\tilde{b} + 120) \right] \tag{A.3.15}$$

$$\mathcal{I}_6(a,b) = \int_a^b du \frac{u^6}{\exp(u) - 1} = \sum_{n=0}^{\infty} \frac{1}{(n+1)^7} \left[\exp(-\tilde{a})(\tilde{a}^6 + 6\tilde{a}^5 + 30\tilde{a}^4 + 120\tilde{a}^3 + 360\tilde{a}^2 + 720\tilde{a} + 720) - \exp(-\tilde{b})(\tilde{b}^6 + 6\tilde{b}^5 + 30\tilde{b}^4 + 120\tilde{b}^3 + 360\tilde{b}^2 + 720\tilde{b} + 720) \right] \tag{A.3.16}$$

A APPENDIX

$$\mathcal{I}_7(a,b) = \int_a^b du \frac{u^7}{\exp(u) - 1} = \sum_{n=0}^{\infty} \frac{1}{(n+1)^8} \left[\exp(-\tilde{a}) \left(\tilde{a}^7 + 7\tilde{a}^6 + 42\tilde{a}^5 + 210\tilde{a}^4 \right. \right.$$
$$\left. + 840\tilde{a}^3 + 2520\tilde{a}^2 + 5040\tilde{a} + 5040 \right) - \exp(-\tilde{b}) \left(\tilde{b}^7 + 7\tilde{b}^6 + 42\tilde{b}^5 \right.$$
$$\left. \left. + 210\tilde{b}^4 + 840\tilde{b}^3 + 2520\tilde{b}^2 + 5040\tilde{b} + 5040 \right) \right], \quad (A.3.17)$$

Please refer to section (7) additionally.

List of Figures

1	Principle of an indirect driven inertial confinement facility.	5
2	Principle of a direct driven inertial confinement facility.	6
3	Definition of the specific intensity.	10
4	Internal energy of a deuterium-tritium plasma depending from temperature.	19
5	Internal energy vs. pressure in a deuterium-tritium plasma at different densities	20
6	Behaviour of the adiabatic coefficients depending from temperature and pressure ratio.	21
7	Cross section of the D-T reaction	24
8	Averaged ionisation stage for a hydrogen plasma depending on temperature and density.	34
9	Ionisation stages of uranium.	38
10	Participation probability of r-times ionised uranium atoms on the ionisation equilibrium	39
11	Ionisation energies of plutonium	40
12	Contributions to the absorption cross section.	43
13	Rosseland averaged mean free path in a D-T and uranium plasma considering free-free transitions only.	50
14	Scattering absorption cross section in a D-T plasma.	51
15	Mean free path in a D-T and uranium plasma including scattering contributions	52
16	Heat flow model.	53
17	Coulomb Logarithm	55
18	Thermal flux limiter	56
19	Heat conduction coefficient	60
20	Thermal diffusivity	61
21	The Compton effect.	72
22	Inscattering contribution	74
23	Layout of the physical models	83
24	Comparison of calculations with and without pure heat conduction by using an eos with $\gamma = 5/3$ and taking the equilibrium radiation contribution into account.	88
25	Results of a calculation with $\gamma = 1.4$. No fusion energy is deposited in the physical system.	93

LIST OF TABLES

26	Snapshots of the Rosseland and Planck mean free path.	94
27	Forming of the hot spot without conduction.	94
28	Results of a calculation with $\gamma = 5/3$. No fusion energy is deposited in the system.	95
29	Results of a calculation with $\gamma = 1.4$. No fusion energy is deposited in the physical system.	96
30	Results of a calculation with $\gamma = 5/3$ taking radiation conduction into account. No fusion energy is deposited in the physical system.	97
31	Results of a calculation with $\gamma = 1.4$. Fusion energy is deposited in the system.	98
32	Results of a calculation with $\gamma = 5/3$. The system is supplied with fusion energy.	99
33	Results of a calculation with $\gamma = 5/3$. Fusion energy is deposited in the system.	100
34	Schematic view of the program system	102

List of Tables

1	Specific heats at constant volume for a D-T plasma.	16
2	Parameters being used for functional approximation of fusion cross section.	24
3	Effective quantum number.	28
4	Electron configuration of uranium by the model of Slater.	28
5	Electron configuration of plutonium by the model of Slater.	28
6	Electron configuration by the orbital model for uranium and plutonium.	29
7	Ionisation energies of uranium by using the atomic model of Slater	30
8	Ionisation energies of plutonium by using the atomic model of Slater	31
9	κ_0^{th} and δ tabulated for hydrogen, uranium and plutonium.	54
10	Photon energy groups.	64
11	Six-group Planck spectrum	66
12	Six-group Planck mean absorption cross section for uranium	67
13	Six-group Planck mean pure free-free absorption cross section for uranium	67
14	Six-group Planck mean Compton cross section	68
15	Analytical evaluated Six-group Planck mean Compton cross section in small photon energy limit.	69

List of Tables

16	Accuracy of the multigroup Compton cross section in small photon energy limit.	69
17	Highest and smallest photon energies $h\nu'$ from which photons are able to scatter into group g.	73
18	Zeroth order moments of the six-group Planck weighted Compton scattering transfer cross section.	79
19	Analytical evaluated Six-group Planck weighted moments of zeroth Legendre order of the Compton scattering transfer cross section in the low photon limit	79
20	Accuracy of the multigroup Compton transfer cross section to zeroth order in the small photon energy limit.	79
21	First order moments of the six-group Planck weighted Compton scattering transfer cross section.	80
22	Analytical evaluated Six-group Planck weighted moments of first Legendre order of the Compton scattering transfer cross section in low photon limit .	80
23	Accuracy of the multigroup Compton transfer cross section to first order in the small photon energy limit.	80
24	Second order moments of the six-group Planck weighted Compton scattering transfer cross section.	81
25	Analytical evaluated Six-group Planck weighted moments of second Legendre order to the Compton scattering transfer cross section in low photon limit	81
26	Accuracy of the multigroup Compton transfer cross section of second order in the small photon energy limit.	81
27	Thermonuclear burn-up in a configuration without α-heating.	85
28	Mass averaged burn-up in calculations without α-heating.	85
29	Thermonuclear burn-up in a configuration taking α-heating into account.	89
30	Mass averaged burn-up in calculations without α-heating.	89

List of Tables

References

[Allen64]   C. W. Allen, *Astrophysical Quantities* (Athlone, London, 1964).

[Apruzese02]   J. P. Apruzese, J. Davis, K. G. Whitney, J. W. Thornhill, P. C. Kepple, R. W. Clark, C. Deeney, C. A. Coverdale and T. W. L. Sanford, *The physics of radiation transport in dense plasmas*, Physics of Plasmas **9** (2002) 2411.

[Armstrong72]   B. Armstrong, *Emission, Absorption and Transfer of Radiation in heated Atmospheres* (Pergamon Press, 1972).

[Atzeni87]   S. Atzeni, *The Physical basis for Numerical Fluid Simulations in Laser Fusion*, Plasma Physics and Controlled Fusion **29** (1987) 1535.

[Atzeni04]   S. Atzeni and J. Meyer-ter Vehn, *The Physics of Inertial Fusion* (Clarendon Press-Oxford, 2004).

[Bell70]   G. I. Bell and S. Glasstone, *Nuclear Reactor Theory* (Van Nostrand Reinhold Company, New York, 1970).

[Bigot06]   D. Bigot, *Inertial fusion science in europe*, Journal de Physique IV **133** (2006) 3.

[Blenski88]   T. Blenski and J. Ligou, *An improved shooting method for one-dimensional schrödinger equation*, Computer Physics Communications **50** (1988) 303.

[Blenski90]   T. Blenski and J. Ligou, *Calculations of radiation opacity for high z elements*, Laser and Particle Beams **8** (1990) 265.

[Brytschkow92]   J. A. Brytschkow, O. I. Maritschew and A. P. Prudnikov, *Tabellen unbestimmter Integrale* (Harri Deutsch, 1992).

[Chan78]   C. Chan, *STEALTH - a Lagrange explicit finite-difference code for solids, structural and thermohydraulic analysis*, Technical Report NP-176-1, EPRI (1978).

REFERENCES

[Chandrasekhar60] S. Chandrasekhar, *Radiative Transfer* (Dover Publications, Inc., New York, 1960).

[Cox68a] J. P. Cox and R. T. Giuli, *Application to Stars*, Vol. 2 of *Principles of Stellar Structure* (Gordon and Breach, 1968).

[Cox68b] J. P. Cox and R. T. Giuli, *Physical Principles*, Vol. 1 of *Principles of Stellar Structure* (Gordon and Breach, 1968).

[Duderstadt82] J. J. Duderstadt and G. A. Moses, *Inertial Confinement Fusion* (John Wiley & Sons, 1982).

[Duffy91] P. Duffy, M. Klapisch, J. Bauche and C. Bauche-Arnoult, *Monte carlo simulation of complex spectra for opacity calculations*, Physical Review A **44** (1991) 5715.

[Ecker63] G. Ecker and W. Kroll, *Lowering of the ionization energy for a plasma in thermodynamic equilibrium*, Physics of Fluids **6** (1963) 62.

[Fiedler07a] J. Fiedler, *Anfänge in der Bestimmung der Ionisierungsenergien und Strahlentransportkoeffizienten in heißen und dichten Plasmen, Teil A*, Technical Report 24, Fraunhofer Institut für Naturwissenschaftlich- Technische Trendanalysen Euskirchen (2007).

[Fiedler07b] J. Fiedler, *Anfänge in der Bestimmung der Ionisierungsenergien und Strahlentransportkoeffizienten in heißen und dichten Plasmen, Teil B*, Technical Report 25, Fraunhofer Institut für Naturwissenschaftlich- Technische Trendanalysen Euskirchen (2007).

[Gonzalez97] J. L. M. Q. Gonzalez and D. Thompson, *Getting started with Numerov's method*, Computer in Physics **11** (1997) 514.

[Hafner91] P. Hafner, G. Locke, J. Schulze and P. Thesing, *Verdichtung und Reaktivitätsaufbau bei einer kugelsymmetrischen Kernspaltungsanordnung*, Technical Report 139, Fraunhofer Institut für Naturwissenschaftlich- Technische Trendanalysen Euskirchen (1991).

References

[Hafner09]   P. Hafner, *Private communication* (2006–2009).

[Heitler57]   W. Heitler, *The Quantum Theory of Radiation* (Oxford University Press, 1957).

[Karzas61]   W. J. Karzas and R. Latter, *Electron Radiative Transitions in a Coulomb Field.*, Astrophysical Journal Supplement Series **6** (1961) 167.

[Khalfaoui97]   A. H. Khalfaoui and D. Bennaceur, *Radiation transport coefficients for degenerate plasma*, Physics of Plasmas **4** (1997) 4409.

[Kourganoff63]   V. Kourganoff, *Basic Methods in Transfer Problems* (Dover Publications, Inc., New York, 1963).

[Kunc92]   J. A. Kunc and W. H. Soon, *Maximum principal quantum numbers of the atomic hydrogen in the solar chromosphere and photosphere*, Astrophysical Journal **396** (1992) 364.

[Landau87]   L. D. Landau and E. Lifschitz, *Statistische Physik*, Vol. 5 of *Lehrbuch der theoretischen Physik* (Akademie Verlag Berlin, 1987).

[Leuthäuser68]   K.-D. Leuthäuser, *Thermodynamische Eigenschaften von Schwermetallen bei hohen Temperaturen*, Technical Report S-2, Fraunhofer Institut für Naturwissenschaftlich- Technische Trendanalysen Euskirchen (1968).

[Levermore79]   C. D. Levermore, *A Chapman-Enskog approach to flux-limited diffusion theory*, Technical Report, University of California, Lawrence Livermore Laboratory Livermore, California 94550 (1979).

[Li09]   J. Li and J. Li, *Angular anisotropy of group averaged absorption coefficient and its effect on the behavior of diffusion approach in radiative transfer*, Journal of Quantitative Spectroscopy and Radiative Transfer **110** (2009) 293 .

[Marshak58]   R. E. Marshak, *Effect of radiation on shock wave behaviour*, Physics of Fluids **1** (1958) 24.

References

[Mayer47]   H. Mayer, *Methods of Opacity Calculations*, Technical Report LA-647, Los Alamos Scientific Laboratory, Los Alamos, New Mexico (1947).

[McCrory08]   R. L. McCrory, D. D. Meyerhofer, R. Betti, R. S. Craxton, J. A. Delettrez, D. H. Edgell, V. Y. Glebov, V. N. Goncharov, D. R. Harding, D. W. Jacobs-Perkins, J. P. Knauer, F. J. Marshall, P. W. McKenty, P. B. Radha, S. P. Regan, T. C. Sangster, W. Seka, R. W. Short, S. Skupsky, V. A. Smalyuk, J. M. Soures, C. Stoeckl, B. Yaakobi, D. Shvarts, J. A. Frenje, C. K. Li, R. D. Petrasso and F. H. Seguin, *Progress in direct-drive inertial confinement fusion*, Physics of Plasmas **15** (2008) 055503 (pages 8).

[Mihalas78]   D. Mihalas, *Stellar Atmospheres* (Chicago Press, 1978).

[Morel00]   J. Morel, *Diffusion-limit asymptotics of the transport equation, the $P_{1/3}$ equations, and two flux-limited diffusions theories*, Journal of Quantitative Spectroscopy and Radiative Transfer **65** (2000) 769.

[Motz79]   H. Motz, *The Physics of Laser Fusion* (Academic Press, 1979).

[Olson00]   G. L. Olson, H. A. Auer and M. L. Hall, *Diffusion, P_1, and other approximate forms of radiation transport*, Journal of Quantitative Spectroscopy and Radiative Transfer **64** (2000) 619.

[Pai66]   S.-I. Pai, *Radiation Gas Dynamics* (Springer, 1966).

[Pakula85]   R. Pakula and R. Sigel, *Self-similar expansion of dense matter due to heat transfer by nonlinear conduction*, Physics of Fluids **28** (1985) 232.

[Pfalzner06]   S. Pfalzner, *An Introduction to Inertial Confinement Fusion*, Series in Plasma Physics (CRC Press, 2006).

[Pomraning73]   G. C. Pomraning, *The Equations of Radiation Hydrodynamics*, Vol. 54 of *International Series of Monographs in Natural*

Philosophy (Pergamon Press, Oxford and New York (reprinted by Dover 2005), 1973).

[Pomraning81]   G. C. Pomraning, *Maximum entropy eddington factors and flux limited diffusion theory*, Journal of Quantitative Spectroscopy and Radiative Transfer **26** (1981) 385.

[Pritzker71]   A. Pritzker, W. Haelg and K. Dressler, *Zustandsgrössen eines schweren Plasmas in Abhängigkeit von Temperatur und Dichte*, Zeitschrift für angewandte Mathematik und Physik **22** (1971) 54.

[Pritzker75]   A. Pritzker, K. Dressler and W. Haelg, *Opacities of high temperature uranium plasma*, Journal of Quantitative Spectroscopy and Radiative Transfer **15** (1975) 1131.

[Pritzker76]   A. Pritzker, W. Haelg and K. Dressler, *Multigroup radiation cross section of high temperature uranium plasmas*, Journal of Quantitative Spectroscopy and Radiative Transfer **16** (1976) 629.

[Pritzker81]   A. Pritzker and W. Haelg, *Radiation dynamics of a nuclear explosion*, Journal of Applied Mathematics and Physics **32** (1981) 1.

[Rutten03]   R. J. Rutten, *Radiative Transfer in Stellar Atmospheres*, 8. Aufl. (Utrecht University lecture notes, 2003).

[Sanchez91]   R. Sanchez and G. C. Pomraning, *A family of flux-limited diffusion theories*, Journal of Quantitative Spectroscopy and Radiative Transfer **45** (1991) 313.

[Simmons00]   K. H. Simmons and D. Mihalas, *A linearized analysis of the modified P_1 equations*, Journal of Quantitative Spectroscopy and Radiative Transfer **66** (2000) 263.

[Slater30]   J. C. Slater, *Atomic shielding constants*, Physical Review **36** (1930) 57.

[Smirnow95]   W. Smirnow, *Lehrgang der höheren Mathematik*, Vol. 3/2 (Harri Deutsch, Frankfurt am Main, 1995).

REFERENCES

[Spitzer53]　　　　　L. Spitzer and R. Härm, *Transport phenomena in a completely ionized gas*, Physical Review **89** (1953) 977.

[Spitzer62]　　　　　L. J. Spitzer, *Physics of Fully Ionized Gases* (John Wiley, Inc., 1962).

[Tahir97]　　　　　N. A. Tahir, L. K.-J., O. Geb and J. A. Maruhn, *Inertial confinement fusion using hohlraum radiation generated by heavy-ion clusters*, Physics of Plasmas **4** (1997) 796.

[Tsakiris87]　　　　　G. D. Tsakiris and K. Eidmann, *An approximate method for calculating planck and rosseland mean opacities in hot, dense plasmas*, Journal of Quantitative Spectroscopy and Radiative Transfer **38** (1987) 353 .

[Vehn97]　　　　　J. Meyer-ter Vehn, *Prospects of inertial confinement fusion*, Plasma Physics and Controlled Fusion **39** (1997) B39.

[Wilson80]　　　　　J. R. Wilson and Lund, C. M., *Some numerical methods for time-dependent multifrequency radiation transport calculations*, Technical Report UCRL-84678, University of California, Lawrence Livermore Laboratory Livermore, California 94550 (1980).

[X-5 Monte Carlo Team03]　X-5 Monte Carlo Team, *MCNP - A General Monte Carlo N-Particle Transport Code, Version 5 - Volume I: Overview and Theory*, Technical Report LA-UR-03-1987, Los Alamos National Laboratory (2003).

[Yan02]　　　　　J. Yan and Z.-Q. Wu, *Theoretical investigation of the increase in the rosseland mean opacity for hot dense mixtures*, Physical Review E **65** (2002) 066401.

[Zel'dovich66]　　　　　Y. B. Zel'dovich and Y. P. Raizer, *Physics of Shock Waves and High- Temperature Hydrodynamic Phenomena*, Vol. 1 (Academic Press, New York (reprinted by Dover 2002), 1966).

[Zeqing06]　　　　　W. Zeqing, P. Jinqiao and A. Jun, *Opacity calculations for high-Z plasma in non-local thermodynamic equilibrium*, Journal of Quantitative Spectroscopy and Radiative Transfer **102** (2006) 402.

[Zimmermann79] G. B. Zimmermann and R. M. More, *Pressure ionisation in laser-fusion target simulation*, Journal of Quantitative Spectroscopy and Radiative Transfer **23** (1979) 517.

REFERENCES

Die VDM Verlagsservicegesellschaft sucht für wissenschaftliche Verlage abgeschlossene und herausragende

Dissertationen, Habilitationen, Diplomarbeiten, Master Theses, Magisterarbeiten usw.

für die kostenlose Publikation als Fachbuch.

Sie verfügen über eine Arbeit, die hohen inhaltlichen und formalen Ansprüchen genügt, und haben Interesse an einer honorarvergüteten Publikation?

Dann senden Sie bitte erste Informationen über sich und Ihre Arbeit per Email an *info@vdm-vsg.de*.

Sie erhalten kurzfristig unser Feedback!

VDM Verlagsservicegesellschaft mbH
Dudweiler Landstr. 99
D - 66123 Saarbrücken
Telefon +49 681 3720 174
Fax +49 681 3720 1749
www.vdm-vsg.de

Die VDM Verlagsservicegesellschaft mbH vertritt

Printed by Books on Demand GmbH, Norderstedt / Germany